The Burning Issues

Er. Mohammad Ashraf Fazili

11/20/2015

The book details some of the burning issues that have caught the attention of The Umited Nations, The Institution of Engineers India and J&K Government, on World Environment Day 2013, 2014 & 2015 Engineers Day 2013, 2014 & 2015, World Water Day -2014, September-14 - Century's worst flood in Kashmir etc.

To My Engineering Fraternity

Contents

01)Theme Write-up of World Environment Day 2013

"Think.Eat.Save"

The theme for this year's World Environment Day celebrations is Think.Eat.Save. Think.Eat.Save is an anti-food waste and food loss campaign that encourages you to reduce your food print. According to the UN Food and Agriculture Organization (FAO), every year 1.3 billion tonnes of food is wasted. This is equivalent to the same amount produced in the whole of sub-Saharan Africa. At the same time, 1 in every 7 people in the world go to bed hungry and more than 20,000 children under the age of 5 die daily from hunger.

Given this enormous imbalance in lifestyles and the resultant devastating effects on the environment, this year's theme – Think.Eat.Save – encourages you to become more aware of the environmental impact of the food choices you make and empowers you to make informed decisions.

While the planet is struggling to provide us with enough resources to sustain its 7 billion people (growing to 9 billion by 2050), FAO estimates that a third of global food production is either wasted or lost. Food waste is an enormous drain on natural resources and a contributor to negative environmental impacts.

This year's campaign rallies you to take action from your home and then witness the power of collective decisions you and others have made to reduce food waste, save money, minimize the environmental impact of food production and force food production processes to become more efficient.

If food is wasted, it means that all the resources and inputs used in the production of all the food are also lost. For example, it takes about 1,000 liters of water to produce 1 liter of milk and about 16,000 liters goes into a cow's food to make a hamburger. The resulting greenhouse gas emissions from the cows themselves, and throughout the food supply chain, all end up in vain when we waste food. In fact, the global food production occupies 25% of all habitable land and is responsible for 70% of fresh water consumption, 80% of deforestation, and 30% of greenhouse gas emissions. It is the largest single driver of biodiversity loss and land-use change.

Making informed decision therefore means, for example, that you purposefully select foods that have less of an environmental impact, such as organic foods that do not use chemicals in the production process. Choosing to buy locally can also mean that foods are not flown halfway across the world and therefore limit emissions.

So **think** before you **eat** and help **save** our environment!

Man is the noblest of all creations. All nature including sun, moon, stars and sky has been subservient to man. Human race is raised up as the best community for mankind. Despite this lofty station, human nature is described as frail and faltering. Whereas everything in universe has a limited nature and every creature recognizes its limitations and insufficiency, man is self destroyed, ungrateful, rebellious, mischief maker, full of pride, arrogating to himself the attributes of self sufficiency.

MAN-THE CROWN OF CREATION

Out of the whole creation on Earth, it is the human being only, who is crowned with the power of a higher intellect and is hence known as 'Ashraf-ul-Makhlooqat'- the crown of creation. He can select what is right or wrong for himself and for his fellow beings and can adopt measures for saving his environment. He can select which food is beneficial for him without making any waste and he can plan, how to save and distribute the surplus food among his fellow creatures that are starving due to unequal distribution or due to poverty.

The other side of the coin is that in spite of being equipped with higher intellect, it is man only who is responsible for generating tremendous waste and all other animals like street dogs, monkeys, wandering cattle, flying birds and even insects serve as scavengers to consume most of this consumable waste and they at times fall prey to the hazardous effects of harmful wastes. Even the trees which are axed by man, exhale oxygen which is inhaled by man and they in turn inhale the carbon dioxide exhaled by man. Besides they bear fruits and fuel for man. The flora and fauna yield fragrance to the atmosphere and add different colors like lavenders, roses, daffodils and tulips etc. to the surroundings. The chirping birds and running streams usher into a music that sends a man to sleep. The singing bees provide curing sweet honey. The medicinal plants provide us with curing drugs. But it is man who pollutes his own atmosphere with carbon dioxide, poisonous gases and causes noise pollution. Man's arrogance has made inroads even into the depletion of his protective ozone layer. He pollutes his own waters of springs, rivers and lakes, which ultimately terminate into the seas of the globe adversely affecting the marine life. He also is responsible for the pollution of soil, which feeds him, sustains him and also which gets trampled under his feet without registering any protest.

EARTH THE ONLY KNOWN LIVING PLANET

The evolution of earth from cold stellar debris to spinning dynamo and the gradual building of the atmosphere may be unique in the universe. No other planet in the solar system at least shows any sign of life. Of the four building blocks for life, hydrogen was created in the first second of the universe, carbon; nitrogen and oxygen were fused in the great nuclear core of a massive star that exploded before our sun was formed. It was the energy of our sun, however, that forged these elements into the complex molecules from which life developed. But, as well as being a giver of life, the sun can also be deadly and the earth requires protection from its lethal rays.

Out of all the galaxies of innumerable existing known stars in the Universe, it is the Earth only which carries life on it, that was considered to be sustained by water, fire, soil and air - (Aab-o-Atash, Khak-o-Baad- the char anasir)-the four basic elements. When life began, 3.5 billion years ago, the first oceans were still only a few degrees below boiling point. There, in waters bombarded by ultraviolet light from space and played on by lightning from massive thunderstorms, the first amino and nucleic acids, the building blocks for life, were formed. It has happened many times and in many places in the Universe-astronomers have detected amino acids floating between the stars- but on earth it gave rise to a history of life which may be unique in its extraordinary variety and abundance.

There are between 3 to 10 million different kinds of plants and animals alive today and possibly twice as many were once alive but are now extinct. Everyone has been built from these basic materials and yet everyone is unique. That is the double miracle of creation: the flow of genetic information across the ages and the editing of that information into almost countless separate categories. The editor is the environment. It selects those organisms for survival that are best suited to life in each particular environmental niche and ruthlessly destroys those that cannot make use of the resources around them.

The story of mankind begins in the tropical forests of at least 65 million years ago. Some 35,000 years ago, human beings physically just like ourselves were living in Europe, Africa and Asia. Over the first 2 million years of human history, our ancestors were entirely dependent upon nature's whims of food. It is only within last 10,000 years that they have settled down to farm the land and control their own food supply.

About 7 billion people are alive today (growing to 9 billion by 1950). Every second three more are added to the total, a growth of more than 10.000 an hour and over 80 million in the space of a year.

Food is mankind's raw energy resource- the fuel that fires the human boiler- and maintaining supplies constitutes man's biggest single concern.

Agricultural efficiency has increased at a staggering pace: in 1980 the World's farms produced twice as much food as they did in 1950. As a result, the earth today grows enough food to support its population, with plenty to spare. But the pattern of production is uneven and many areas still go short.

Thousands of different kinds of plants are consumed by man but first three – wheat, corn and rice –account for about half of the world's harvest. By no means every niche of the planet's surface can be exploited for crop farming, however for a combination of three basic factors- sunshine, moisture and soil – determines where the global harvest can be gathered in. At present only 11 percent of the earth's land surface is farmed for crops, while a further 20 percent is thought to be cultivable.

GREEN REVOLUTION

One major factor in the recent boon in food production has been the development of new, high-yielding strains of wheat, corn and rice. Cultivated with modern fertilizers, herbicides, pesticides and irrigation techniques, these grains have generated what is known as the Green Revolution. Bumper harvests have been the result. But the techniques modeled on the practices of U.S agriculture have had their critics too. The chemicals required by high-yield strains are derived chiefly from fossil fuels, which have become increasingly costly since the oil crises of 1973. Mechanized farming too, consumes energy resources- about 25 gallons of gasoline are required to produce one acre of corn in USA. Such farming tends to benefit large farmer with capital to invest at the expense of the small farmer.

The world is witnessing adjustment towards the kind of organic farming practiced in China. Here crop waste is recycled to provide fertilizer, so that less synthetic matter is required. In addition mixed cropping is practiced: grains are planted with legumes- the soybean- is already a post world war 2nd success story in the developing world. It is grown increasingly for its high protein content and adaptability and its oil is used for making paints and chemicals, as well as margarines and cooking oils.

The greatest hopes for feeding future generations lie in plant breeding and genetics. Resistance to pests and diseases can be bred into crops so that spraying with hazardous chemicals becomes increasingly obsolete. Strains may be developed to cope with harsh climates of desert or tundra. Modern techniques of gene transfer offer possibilities for cultivating radically improved species. Tens of thousands of potentially edible species have been identified and it may yet prove possible to carpet the world's most barren wastes with new forms of nutritious vegetation.

HOW THE PLANET PROVIDES

Prophet Muhammad (PBH) said over 1400 years back: Planting a tree is a continuous source of blessing (Sadaqa Jariah) for the person who plants it. Shaikh-ul-Alam,

Shaikh Noor-ud-Din Noorani (RA) (d. 1430 AD) – a great Kashmiri Sufi saint said about six hundred years back- "ANN POSHI TELI YELI WAN POSHI"- i.e. food is subservient to forests.

The dense multilayered canopy of the tropical forests is one of the miracles of life on earth. It covers only 8 percent of earth's surface but may harbor more than 40 percent of all species of plants and animals. This massive biological resource not only provides fuel and building materials but is a critical regulator of the world environment, protecting the topsoil, governing even the flow of rivers, and on a global scale, helping to maintain a balanced climate. But the forests are under pressure. As many as 100 acres of tropical forests are being destroyed every minute- a total of over 83,000 sq. miles a year. Perhaps twice that area is being seriously degraded. Timber harvesting and cattle ranching play their part in the destruction, but as significant are the slash and burn techniques of the 150 million forest farmers in the world today. Some places such as westernmost Amazon and parts of the Zaire basin, where the population is not so intense can be sure to survive at least partly intact.. But in western Africa, Southeast Asia, and the Himalayan watershed, most of the unique and the irreplaceable forest will soon have gone forever. The only real solutions to the destruction of the tropical rainforest lie in the control of population growth; increasing yields of good land; planting alternative sources of timber and fuel wood; and ecological sensitivity in development projects.

DESERTIFICATION

About one third of earth's surface, fringing the great deserts, is arid, often subjected to prolonged draught. Natural ecosystems here are adapted to long dry spells and can take them in their stride. But when such areas are put under pressure by rising populations and the demand for food, their ability to spring back to productiveness after draught is severely reduced. The rains when they come, rather than soaking into and feeding the land, can wash away the soil, extending the limits of the desert. Every year about 47,000 sq. miles of agricultural land are made worthless this way. The process of desertification is probably worsened by the increasing levels of carbon dioxide and other gases in the atmosphere, produced by the burning of fossil fuels, which are likely to raise global temperature within this century by up to 4.5 degrees C. A problem of such vast proportions is not easily dealt with. A reduction in carbon dioxide emission from the industrial countries will help. Keeping animals and vehicles off certain areas can reestablish vegetation. The deliberate planting of forest and shrub stands- in the Soviet Republics, India, and China – reduces erosion on a local scale.

OUR FOOD

Climate and environment are the world's great chefs, giving Mexico its tortillas, Greece its goats milk, cheese; China its pork spareribs and Japan its seafood dishes. And it is regional variations in these two factors that strongly influence what is raised where.

As stated above the world's three main cereals are wheat, corn and rice, each of which has its special needs. Wheat is a crop of the temperate prairies and will tolerate very cold winters. Corn is vulnerable to frost and is therefore confined to a warmer climate band. And rice favors the special combination of warmth and copious rainfall that is found especially in monsoon zones.

Grain constitutes about half of the world's food production by weight, but similar factors associate other crops with particular environments: for example, grapes with Mediterranean climates and the potato with dull, cloudy skies and clammy soils.

There are vast expanses of desert and bleak uplands whose lean and rocky soils support little more than coarse grasses. Since the human stomach cannot digest grass, it is the livestock here in particular the sheep and the goats- that act as our food converters, yielding meat, milk and cheese.

Cattle can be raised in a temperate band stretching from the edge of the Sahara to the margins of the Arctic Circle. But cattle, like sheep are ruminant's digestive system calls for a diet chiefly of grass and which require wide grazing area. These are an inefficient food resource for the world's overpopulated regions and due to vulnerability to the tsetse fly, are especially scarce in the humid tropics. China is the main producer of pork yielding nearly 40 percent of the global total.

Fish like all other food-stuffs, display preferences for habitat. Cod favors the cold waters of the North Atlantic, while tuna prefer warmer seas; flatfish, such as halibut feed on the seabed, while herring cruise close to the surface. The principal fishing grounds are all in coastal zones where nutrients, leached from the land, mix with the rich sediment that is swept up from the sea floor by ocean currents and offshore winds. These waters comprise our teeming marine meadow lands, thick with tiny plankton supporting larger organisms that are, in turn, consumed by shoaling fish. In total the earth's fishing fleets bring in some 68 million tons a year. Japan, with its intricate network of islands, has an ancient fishing tradition and remains the largest single harvester of the sea.

AVAILABILITY AND FAMINES

If the global harvests were shared out equally, each person could receive 5lbs. (2.3 kgs.) of food per day. Hunger need never be with us.

The reason why famines still take their terrible toll has more to do with the complexities of politics, economics, storage and distribution than with the physical capacity of the earth itself. The planet is fertile. Science has opened up new possibilities. And, in the opinion of many experts the age old scourge of hunger could with global cooperation, be eradicated in the near future.

To meet future needs, we can colonize the world's inhabitable areas. The earth's total cultivable land is some 7.9 billion acres, of which less than half is currently being farmed. Although the remainder may be harsh or inaccessible terrain, we have the means to drain swamps, plant hillsides and bring deserts into bloom.

One short term response to starvation in the third world is to transport surplus food from where it is stockpiled to where it is needed. The biggest grain exporters are USA, Canada, Australia and Argentina. Thanks to the green revolution, India, Thailand, Burma and Surinam can now be added to the list of smaller net exporters. Many others for example Mexico and the Soviet Republics would be the net grain exporters but for the demands of live stock, which now consume more grain than grass.

Lakhs of animals are slaughtered on *Id-uz-zuha* in *Haj* pilgrimage, only a small part of which would be distributed in earlier days and the rest bull-dozed into the ground, but now it has been made possible to dispatch the surplus meat to starving countries.

In the long term, however the transporting of surplus food does nothing to help farmers in poor countries to produce more. Indeed, pouring cheap food into third world can lower prices there so much that local farmers are put out of business. Except in emergencies, perhaps what poor countries need most is appropriate technology, transport facilities, education and better administration.

One global measure of food production is provided by the average number of calories supplied by the agriculture of different countries. How many calories an individual actually needs depend on his or her body weight, type of activity and the environmental temperature. Accounting for these variables, the FAO of UN estimates the average daily needs of a person in Finland, where a relatively old population lives in a tropical climate; the average is 2,160 calories per day. One must eat what he needs and overeating leads to obesity and other resultant diseases, besides wastage of food.

Wastage of food is prohibited by all. The Holy Quran says: "KULU WASHRABU WALA TUSRIFU"- You may eat and drink but do not cross the limits. A Hakim from Syria stayed for six months in Madina at the time of Prophet Muhammad (PBH), but had to leave as no patient visited him. On enquiry, he was told that people fill a third of their stomach with solid food and a third with liquid and leave a third empty. The Hakim concluded that most of the diseases are stomach related, hence the result. Besides there are many instances, when people served the needy and themselves preferred to go hungry.

Here is a lesson for us not to serve excessive food with meat in *wazwan* resulting into waste that could feed many more starving people. In olden days, sharing food on a plate by four people served with just seven preparations of meat/vegetables would generate no wastage, but the later extravagance and false showmanship with over 20 preparations of meat and chicken has made it a curse in these hard times. The height of things is that we indulge and participate with great interest in these extravagant functions yet simultaneously lamenting and condemning while sharing the *wazwan*. In this behalf the procedure adopted by Arabs and South Indians is preferable, when they gather around a huge plate full of *Biryani* etc, and pour their desirable share in their respective plates causing zero wastage. Many people are shifting to buffet service now in Kashmir too which eliminates wastage.

J&K – FOOD DEFICIT STATE:

The Economic Survey 2012-13 reveals that J&K State is a food deficit and consumer state. The state is mostly depended on the import of food grains from other states and that the day-by-day dependence of the state for food grains from outside is increasing. During 2011-12, import and off take of food grains stood at 908.22 and 856.27 thousand metric tones which respectively were 20 per cent and 14 percent more than the previous year.

LAND CONVERSION:

It is also reported that due to random conversion of farm land, the agricultural production is decreasing day by day. While the State Government in 2011 had decided to bring law to ban the conversion of farm land for commercial or non-farm use, the law is yet to see the light of the day. In Kashmir alone more than two lakh kanals of agricultural land of the net sown or cultivated area of 3.5 lakh hectares has been converted for commercial and other purposes. The situation is no better in Jammu. The conversion of farmlands for residential purposes has negative consequences on food security, water supply besides health of the people, both in the cities and in the peripheries. If the conversion of agricultural land continues at the same pace, in coming years the state would have no agricultural land according to an official of SKUAST. Agriculture was the largest land use category in 1971, covering an area of 70.13 per cent of Srinagar district which has been reduced to 48.93 percent in 2009 at an annual rate of -0.79 percent.

The maximum conversion has taken place along major transportation corridors around the city, therefore giving rise to ribbon settlements. Presently dozens of colonies are coming up on agriculture land in different parts of the Valley. Even residential houses and restaurants are being constructed on it. The law enforcement agencies need to curb the menace before the problem assumes horrendous proportions."

FOOD SECURITY BILL

A healthy development has been that the Food Security Bill is being introduced in the current session of Parliament.

The Bill seeks "to provide for food and nutritional security in human life cycle approach, by ensuring access to adequate quantity of quality food at affordable prices to people to live a life with dignity and for matters connected therewith and incidental thereto".

It extends to the whole of India and "shall come into force on such date as the Central Government may, by notification in the Official Gazette appoint, and different dates may be appointed for different States and different provisions of this Act"

The Bill has three schedules (these can be amended "by notification"). Schedule 1 prescribes issue prices for the PDS. Schedule 2 prescribes "nutritional standards" for midday meals, take-home rations and related entitlements. For instance, take-home rations for children aged 6 months to 3 years should provide at least 500 calories and 12-15 grams of protein. Schedule 3 lists various "provisions for advancing food security", under three broad headings: (1) revitalization of agriculture (e.g. agrarian reforms, research and development, remunerative prices), (2) procurement, storage and movement of food grains (e.g. decentralized procurement), and (3) other provisions (e.g. drinking water, sanitation, health care, and "adequate pensions" for "senior citizens, persons with disability and single women").

J&K State has been the pioneer state in providing food rations to urban areas right from Maharaja's time, which is/was followed by other states.

A LESSON TO LEARN

We have a lesson to learn from the newly developing cities of Abu Dhabi and Dubai. These cities are littered with no solid/liquid wastes or polythene; have no noise pollution or dust pollution indoors, no street dogs, no beggars, no uniformed/armed policemen to be sighted. Building laws are strictly followed and every spot is designed and executed as per consultant's advice with all aesthetic appeal. The cities have developed with 5-lane roads either way with foot paths bedecked with green grass and flowers and palm trees with intermittent irrigation facilities. During the past about four decades, mud huts have got converted to sky scrapers including the world's tallest building Burji Khalifa. Metro has been introduced for a faster travel. A new city "*Masdar*" is coming up all based on solar energy banning use of fossil fuels in all devices.

INDIVIDUAL EFFORTS NEEDED

In our homes the compostable solid wastes can be converted into useful compost for kitchen garden by dumping it regularly in a circular pit of one meter diameter and one and a half meter depth, with no lining but covered with a lid. Addition of lime over every foot layer is advisable. This could eliminate the use of harmful chemical compost. Individual efforts can make a big impact and reduce the burden on the Municipal authorities, who have utterly failed to establish a mechanical compost plant in Srinagar city, recommended by UEED three decades back.

Similarly construction of a cheaper alternative of twin pit low cost *'suchalya'* instead of the costly sanitary fitted toilets with septic tank and soakage pit has proved a success story in many cities/ villages of India and in more than 21 UNDP countries. Our rural/urban sanitation needs to adopt this program in a big way.

02) FRUGAL ENGINEERING - ACHIEVING MORE WITH FEWER RESOURCES

The Oxford Dictionary defines Engineering as: -the branch of science and technology, concerned with the design, building, and use of engines, machines, and structures.

-a field of study or activity concerned with modification or development in a particular area: *software engineering*

-the action of working artfully to bring something about.

The aim of engineer is to make use of the material economically getting maximum benefits within the prescribed limit of factors of safety as per BSI code of practice.

Frugal is defined as:
-sparing or economical with regard to money or food.

Thus the term frugality is already inscribed in the term "engineering" and 'frugal engineering' is to be super-economical within safe limits.

Frugal Engineering is the science of breaking up complex engineering processes into its basic components and then re-building each component in the most economical manner. The end result is a simpler, more robust and easier to handle final process. It also results in a much cheaper final product which does the same job qualitatively and quantitatively as a more expensive complexly engineered product.

It is generally believed that Indians and other South Asians are the most adept in frugal engineering, because resources and capital are scarce in this region.

Many terms are used to refer to the concept. "Frugal engineering" which was coined by Carlos Ghosn, the joint chief of Renault and Nissan, who stated, "frugal engineering is achieving more with fewer resources."

In India, the words "Gandhian" or "*jugaad*", Hindi for a stop-gap solution, are sometimes used instead of "frugal". Other terms with allied meanings include "inclusive innovation", "catalytic innovation", "reverse innovation", and "BOP innovation", etc.

At times this no frills approach can be a kind of disruptive innovation.

History

Spotlighted in a 2010 article in *The Economist*,[1] the roots of this concept may lie in the appropriate technology movement of the 1950s although profits may have been first wrung from underserved consumers in the 1980s when multinational companies like Unilever began selling single-use-sized toiletries in developing countries. Frugal innovation today isn't solely the domain of large multinational corporations; however, as small, local firms have themselves chalked up a number of homegrown solutions. While General Electric may win plaudits for its US$800 EKG machines, cheap cell phones made by local, no-name companies and prosthetic legs fashioned from irrigation piping are also examples of frugal innovation.

The concept has gained popularity in the South Asian region, particularly in India. The US Department of

Commerce has singled out this nation for its innovative achievements saying in 2012 that "there are many Indian firms that have learned to conduct R&D in highly resource-constrained environments and who have found ways to use locally appropriate technology.

Notable innovations

Frugal innovation is not limited to durable goods such as the GE US$800 EKG machine or the US$100 One Laptop per Child but also services such as 1-cent-per-minute phone calls, mobile banking, off-grid electricity, and microfinance.

ChotuKool fridge

A tiny refrigerator sold by Indian company Godrej, the ChotuKool may have more in common with computer cooling systems than other refrigerators; it eschews the traditional compressor for a computer fan. (It may exploit the thermoelectric effect.)

Jaipur leg

A low cost prosthetic developed in India, the Jaipur leg costs about $150 to manufacturer and includes some clever improvisations such as incorporating irrigation piping into the design to lower costs.

Mobile banking

Mobile banking solutions in Africa, like Safaricom's M-Pesa, allow people access to basic banking services from their mobile phones. Money transfers done through mobiles are also much cheaper than using a traditional method. While some services can be accessed on a mobile alone, deposits and withdrawals necessitate a trip to a local agent.

Nokia 1100

Designed for developing countries, the Nokia 1100 is basic, durable, and–besides a flashlight–has few features other than voice and text. Selling more than 200 million units only four years after its 2003 introduction, has made it the best selling phone of all time.

Solar light bulb

In some Philippine slums, solar skylights made from one liter soda bottles filled with water and bleach provide light equivalent to that produced by a 55 watt bulb and may reduce electricity bills by US$10 per month.

Tata Nano

Designed to appeal to the many Indians who drive motorcycles, the Tata Nano was developed by Indian conglomerate Tata Group and is the cheapest car in the world.

The Importance of Frugal Engineering

Frugal Engineering will be of great relevance to developing countries, as a flexible approach that perceives resource constraints as a growth opportunity. According to Paul Polman, CEO of Unilever, at the current rate of consumption, by 2030 we would need two planets to supply the resources we need and to absorb our waste. As engineers, in the service of the humanity enabling the citizens to enjoy a better quality of life, we have an added responsibility these days to find engineering solutions – of course, frugal - to problems thrown up by all sectors endangering the environment.

Providing new goods and services to "bottom of the pyramid" customers requires a radical rethinking of product development.

A cell phone that makes phone calls — and does little else; a portable refrigerator the size of a small cooler; a car that sells for about US$2,200 (100,000 rupees). These are some of the results of "frugal engineering," a powerful and ultimately essential approach to developing products and services in emerging markets.

To get a handle on what frugal engineering is, it helps to understand what it is not. Frugal engineering is not simply low-cost engineering. It is not a scheme to boost profit margins by squeezing the marrow out of suppliers' bones. It is not simply the latest take on the decades-long focus on cost cutting.

Instead, frugal engineering is an overarching philosophy that enables a true "clean sheet" approach to product development. Cost discipline is an intrinsic part of the process, but rather than simply cutting existing costs, frugal engineering seeks to avoid needless costs in the first place. It recognizes that merely removing features from existing products to sell them cheaper in emerging markets is a losing game. That's because emerging-market customers have unique needs that usually aren't addressed by mature-market products, and because the cost base of developed world products, even when stripped down, remains too high to allow competitive prices and reasonable profits in the developing world.

Frugal engineering recalls an approach common in the early days of U.S. assembly-line manufacturing: Henry Ford's Model T is a prime example. But as industries grew and matured over the decades, and as consumers prospered to levels few would have predicted a century ago, product development processes became hardwired and standard operating procedures worked against frugality.

In addition, the profit structure in mature markets reduced incentives for major change. Constant expansion of features available to consumers in the developed world, frivolous or not, has provided many businesses with their richest profit margins. Mature-market customers continue to accept price premiums for new features, leading companies to over-engineer their product lines — at least from the point of view of emerging-market customers. The virtual extinction of manual car windows in the United States is just one example.

Frugal engineering, by contrast, addresses the billions of consumers at the bottom of the pyramid who are quickly moving out of poverty in China, India, Brazil, and other emerging nations. They are enjoying their first tastes of modern prosperity, and are shopping for the basics, not for fancy features. According to C.K. Prahalad, author of *The Fortune at the Bottom of the Pyramid* (Wharton School Publishing, 2005), these potential customers, "un served or underserved by the large organized private sector, including multinational firms," total 4 to 5 billion of the 6.7 billion people on Earth. Although the purchasing power of any of these new consumers as an individual is only a fraction of a consumer's purchasing power in mature markets, in aggregate they represent a market nearly as large as that of the developed world.

Attracted by the size and rapid growth of emerging markets — concurrent with a growth slowdown in the developed world — companies in a range of industries are establishing distribution and manufacturing operations as well as research and development centers in these regions. However, some of these companies may not fully grasp the challenges that competition in emerging markets entails. The prospect of high-volume profit streams may be enticing, but those profits must be earned in the face of lower prices, lower per-unit profits, and stringent cost targets.

In addition, too few companies realize how demanding emerging-market customers can be. They don't spend easily, because they don't have much to spend. They require a different set of product features and functions than their developed-world counterparts, but still insist on high quality. Global companies, therefore, must change the way they think about product design and engineering. Simply selling the cheapest products on hand or reusing technologies from higher-priced products will not cut costs enough and is unlikely to result in the kind of products these new customers will buy. The central tenet behind every frugal engineering decision is maximizing value to the customer while minimizing nonessential costs. As already stated the term *frugal engineering* was coined in 2006 by Renault Chief Executive Carlos Ghosn to describe the competency of Indian engineers in developing products like Tata Motors' Nano, the pint-sized, low-cost automobile. Companies such as Suzuki paved the way for the development of low-cost automobiles, but there may be no better example of frugal engineering than the Nano, which will allow millions of people with modest means to reliably drive their own car. The Nano is not — like so many other low-cost vehicles — a stripped-down version of a traditional, more expensive car design. Like other newly engineered products selling well in emerging markets, ranging from refrigerators to laptop computers to X-ray machines, it is based on a bottom-up approach to product development.

Even global companies uninterested in the growth offered by the world's lowest-income consumers will have to pay attention to the lessons of frugal engineering: Products developed with this approach are beginning to compete with goods sold in developed countries, a trend that's likely to continue. Deere & Company, for example, designed and sold small, lower-powered tractors in the Indian market, but didn't begin selling such models in the U.S. until an Indian company, Mahindra & Mahindra Ltd., beat them to it. Mahindra & Mahindra has proven an able competitor to Deere in larger tractors as well. General Electric (GE), on the other hand, has been more proactive; for example, it has sold a revolutionary new low-cost handheld ultrasound scanner in developed markets by incorporating frugal engineering lessons learned in its Indian medical research and development lab. A low-cost GE electrocardiogram machine, developed at the same Indian lab for the local markets, is now being sold in the United States and Europe as well.

Meeting all these challenges will require a change in corporate culture. Some companies will be up to it; other companies will not. A successful approach to frugal engineering involves new ways of thinking about customers, innovation, and organization.

Understanding the Customer
The ultimate goal of frugal engineering couldn't be more basic: to provide the essential functions people need — a way to wash clothes, keep food cold, get to work — at a price they can afford. Critical attention to low cost is always accompanied by a commitment to maximizing customer value. The Tata Nano development team's decision not to include a radio on the standard model wasn't a simple move to avoid cost. The team understood that the typical Nano customer places far more value on extra storage space. Using what normally would be the radio slot for storage not only avoided a major cost, but also added value for the customer.

Such carefully calculated trade-offs, made at the product planning stage, serve the dual purpose of maintaining low costs and increasing the product's overall functionality and utility for the buyer. Assessment of those trade-offs requires close, careful observation on the part of planners if they are to arrive at a deep understanding of the ways a product fits (or doesn't fit) into customers' lives.

Again the Nokia 1100 cell phone is another example. Experience has shown that when low-income people in just about any country begin to enjoy a bit of economic prosperity, one of their first purchases is a cell phone. Many new cell phone customers in emerging markets are agricultural workers who spend their days outdoors. When Nokia developers watched field-workers using mobile phones in India, they noticed that the intense humidity made the phones slick and hard to hold or dial. So the phone was built with a nonslip silicon coating on its keypad and sides. The handset was also designed to resist damage from dust that is common in arid climates and some factory environments. The phones are otherwise basic: They can send and receive phone calls and text messages. The screens are monochrome. Because the phones lack fancy software, the power draw is smaller, so they can operate longer between charges. The only real extra is a tiny, energy-efficient flashlight that's proven popular in areas where power blackouts are common — in other words, in most rural villages and many emerging-market cities. At a price of $15 to $20, the Nokia 1100 is the best-selling cell phone ever.

More than a year after coining the term "frugal engineering" to describe Indian engineers, Carlos Ghosn, the joint chief of Renault and Nissan, is still not frugal with his praise for Indian techies.

And his love affair with the country, which isn't exactly globally acclaimed for engineering skills, continues.

Flying in to Chennai, which is fast becoming an auto hub, Ghosn once again recently lavished compliments on engineers.

"Frugal engineering is achieving more with fewer resources. The cost of developing a product in the West is high since engineers there use more expensive tools. In India, they achieve a lot more with fewer resources," Ghosn said.

Between Nissan and Renault, there are now three joint venture companies with Indian partners for different products. Renault and utility vehicle manufacturer Mahindra & Mahindra have a JV to manufacture Logan cars in India.

Renault, Nissan and M&M also have a three-way JV to manufacture cars for the respective principals. Now, Nissan has a JV with Ashok Leyland for the light commercial vehicle (LCV) segment

. "We see a lot of opportunities for LCVs in India but we would not have come alone. We were looking for a partner. India is a sophisticated market that requires sophisticated products and we would have wasted a lot of resources had we tried to come alone,"

Ghosn told media persons here.

There could be more JVs from the group in the future.

Nissan, Renault and Bajaj Auto are in talks to develop and manufacture "ultra low-cost" passenger cars in India.

"We will enter into as many JVs as required," said Ghosn, who flew out to Pune recently to hold interactions with Bajaj Auto officials for the low-cost car.

Ghosn operates out of two continents -Paris in Europe and Tokyo in Asia -and looking at the number of visits he would be making on account of the multiple business interest his group has here, Dheeraj Hinduja, co-chairman, Ashok Leyland, in a lighter note, said that he has a third headquarter in India.

India will be a centre of frugal engineering

RA Mashelkar, former director-general at the Council of Scientific and Industrial Research (CSIR), and national Research professor, has thought long and hard about Gandhian engineering— his version of frugal engineering, the term coined by Nissan CEO Carlos Ghosn. Malshelkar, for his part, became aware of the true extent of the practice in India only when he instituted an award on inclusive innovation in memory of his mother. There were more than a hundred entries for the award that was given on December 17. The two joint winners had developed two low-cost solutions for rural India: a portable device to detect five eye diseases and a diaper that costs one-tenth its current price.

Nature is the best teacher of frugal engineering:

Every creation animate or inanimate is designed by nature with exact specifications, taking an example of atom, its number of electrons, protons, neutrons that determine the individual characteristics of every material. There was once a description of analysis of human body in TOI about half a century back. It quantified the calcium, potassium, magnesium etc contained in the human body, which was priced at Rs. 3.50 only. By this meager amount nature had created an automatic machine that could produce many more of its prototypes. In the end it had concluded that we are simply wonderful. So is the case with other creations right from infinitely vast universe studded with gigantic galaxy of stars, milky ways and black holes, our solar system, our planet earth and all the environment and elements suitable for the sustenance of life on it, to the creation of animals, plants, insects and microbes (not seen by naked eyes).

J&K Scenario

Many innovations have been recently made by the young entrepreneurs in J&K State, but it seems that they lack the support deserved by them to push their innovations ahead into manufacturing stage. The recent one is a joint venture of a professor and a student (as shown in a TV show- Good Morning J&K) invention of a turbine that can generate electricity just on running water without any water-fall, which has a tremendous potential in solving the power crisis of the State. Kashmir University is reportedly helping such innovators to promote their projects.

Similarly a young engineering student of Kashmir, Arif Moosvi developed web designing framework Hotsky, used for developing website. This is India's first web designing framework. Earlier Asif Ahmad- a Kshmiri boy developed an android application- Droid Explorer which was hosted by Google Play. The application has witnessed 5000 downloads worldwide. Recently a 19 year old boy developed an android game based on basic principles of physics. Earlier a 23 year old software engineer developed an android application- "Dial Kashmir" that contains over 500 contacts of Govt. and private departments. Another young engineer developed an online platform where people can share and get any information regarding Kashmir.

Otherwise too, the Kashmiris have been practicing frugal engineering earlier than the advent of machines. With limited available resources, they had devised their own cheap devices like *Wagu-* a grass mat, *Pulhur-* a grass slipper, *khraw-* a wooden slipper, *Tathul-* a wooden tub, a watermill- *(grath)* for grinding maize, wheat and spices which has been in use for centuries together. The *"Yinder"* to spin *Pashmina wool* was a common domestic tool with its accessories. The *"Kanz"and "Muhul"* was used to pound rice, thereby by providing an exercise to our woman-folk. Similarly the copper teapot *"samawar"*-that keeps our tea hot, while we sip it. Then *"kangri"/ "mannan"* –the firepot that kept us warm in severe winters. So were our *'hamams'* that made us face cool temperatures. Our mud hearths *"dhan"* had a water container called *"matti"* attached to it, whereby water would get heated along with the cooking of meals and the residual charcoal would be used in *"kangris".* Similarly the popular dress of *"Pheran"*, *"Tilla work"* had its own charm and utility. Again the *"jajir / hooka"* used for smoking tobacco was also an indigenous innovation. The recipes of the balms prepared by the barbers for treatment of boils, wounds etc. are lost with their deaths. *"Wazwan"* too has its own identity and charm. *Kashur Kagaz-* the kashmiri paper was washable. In construction works *dajji-diwari, panjra-kari, pachar bandi, khutum bandi* etc. was indigenous innovation. *Koshur put- the home spun Kashmiri pattoo, Kashmiri shawl with embroidery, Kani shawl, Pashmina, Shah-tush, Kashmiri silk were all local made. Woollen Namdhas and Ghabbas, Paper machie, silver work, copper work, wood carving, fur making, wooden boats, dungas, house boats, even tongas pulled by horses* have their own individuality. Like that there are many more innovations which have been invented due to the necessity of the times and availability of the limited and scarce resources; confirming the saying that: *"Necessity is the mother of invention".*

Thus a Kashmiri is born with an innovative brain, given the chance and encouragement; he can be a great contributor to the "frugal engineering" even in modern times.

Kashmir has produced many fertile brains in the form of saints, historians, scholars, poets, artists, painters, kings, politicians of international reputation, which include Saga Nila, Kalhana, Abhinavgupta, Nagarjuna, Lalitadattya, Lala Arifa, Shaikh Noor-ud-Din Noorani, Shaikh Yaqub Sarfi, Habba Khatun, Mulla Mohsin Fani, Mulla Tahir Gani Ashai, Akhund Mulla Kamal, Molvi Anwar Shah Kashmiri, Zain-ul-Abideen (Budshah) etc., Even the forefathers of Jawaharlal Nehru hailed from Kashmir.

Here I quote Dr. Iqbal (d. 1938) - the great philosopher poet who too was of Kashmiri origin:

 "Jis khak ke zameer mein ho aatash-i-chinar; mumkin nahin ki sard ho who khak-i-arjamand"

i.e. The dust instinct with the fire of Chinar—That fiery dust will never turn cold.

Again the son of the soil Nunda Reshi (RA)(d. 1439 AD) said:

I broke my sword and made it into a sickle.

03) <u>PROPOSED SKEWED BRIDGE ON RIVER JHELUM and FLY-OVER FROM JEHANGIR CHOWK TO RAMBAGH</u>

(My GK Write up)

The Master Plan of Srinagar Metropolitan Area-2000-2021 highlights the problem in transportation sector being the concentration of activities in the Central Business District (CBD) extending from Dalgate to Batamaloo. All major Government, Commercial and Transport terminals are located in this area. Some broad junctions on Moulana Azad Road like Budshah Chowk, Jehangir Chowk and Batamaloo have heavy peak hour traffic volumes ranging from 1900 to much above 2000 Passenger Car Units (PCU's). Even inter-city traffic passes through this area.

The decentralization of activities is recommended in the Master Plan for improvement in the transportation system. Non-conforming uses in CBD have been identified which are proposed to be removed from the existing sites under the hypothesis of decentralization. In order to deplete the traffic intensity on Moulana Azad Road and its junctions , Dood ganga road from Rambagh to Batamaloo and East West corridors, one on the North from Baba Dharam Das (Kohna Khan) to Neelam Chowk via Barbarshah and Ganpatyar and another from Lasjan to Gangabug via Padshahibagh, Mahjoor Nagar and Rambagh have been proposed. The non-conforming uses in the CBD once shifted will deplete the traffic to a great extent. These include shifting of KMDA Bus Stand, Batamaloo Bus Stand, Matador Stand, Police Lines/PCR at Magarmalbagh/Batamaloo, Jail at Kathidarwaza, Defence use at Tatoo Ground, JKLI at Doodganga behind Shergrhi Police Station, BSF frontier Head Quarters at Sanatnagar, LPG/Petrol Storage at Sanatnagar, Shali Store Complex, Storage Godown at PCR Batamaloo, Farm Land in Agriculture complex Lal Mandi, Lower Courts at Lal Chowk, Govt. Press Lal Chowk, Hoticulture Farm near Convent School, Saw Mills and Timber Depots of the City particularly from

C.B.D., Gwalas in core city, Workshops along major roads at KMD, Bagi Nand Singh, River Bank, Nallah Beds etc., Stone Crushers on N.H.Way from Pampore to Sonawar, Skin Trade, Taxidermy in core city, Leper Asylum, Mental Hospital, Exhibition Complex, CONFED on I.G.Road near Magarmal crossing, Veternary Complex at Maisuma, Forest Deptt building near Lower Court, Other enlisted sordid/slum Pockets, Karamchari Colony of Srgr. Muinicipality at Rambagh and around fort at Kalai-andar, petrol Pumps at Batamaloo and M.A. Bridge, Milk Chilling Plant at Cheshmashahi, Brick kilns at various locations in existing areas etc. as envisaged in the Master Plan. Hence the consequent depletion of traffic obviates the ventures like the Construction of the proposed Convent Bridge and Rambagh Flyover.

Proposed Transportation Plan:

A transportation Plan is envisaged in the Master Plan, wherein a proper hierarchy of streets is proposed. The upper rung is the proposed Motorway with 200 ft. Right of Way starting from Galandar on National Highway 1-A to Narbal Crossing of Srinagar Baramulla road via Nowgam, Humhama and Soibug. All the roads crossing on the proposed motorway shall be on a different grade. The proposed motorway shall have a limited accesses and exits. The total length of the motorway shall be about 28 kms. This motorway shall be an important bye pass route for goods as well as regional traffic. The motorway shall be connected to the CBD through a 90 ft. wide road through Nowgam and Rambagh. The road shall also connect the Railway Station at Nowgam and Freight Centre with the motorway, thereby segregating the goods and the passenger traffic. The transport network also comprises of the arterial roads of 90 ft. to 150 ft. right of way and sub-arterial roads of 68ft. to less than 90 ft. along with North-South and East-West corridors. Out of these corridors some are existing with bottlenecks to be removed and some are proposed afresh. Collector streets from residential neighbor-hoods are planned as 40 ft. to 68 ft. wide with high level of service. All these roads put together constitute about 1753 hectares or 7.35 % of the proposed developed area with the total length of proposed road network above 40 ft. being 506 kms. In practice the road length is not increased by a considerable margin. Therefore it is proposed to increase the Right of Way and carriageways and consequently the level of service.

Grade Separated Intersection:

The traffic capacity of the existing arterial system in Srinagar city has crossed its limits. There is a forced flow on major roads according to the traffic survey. The peak hour volume is more than 3000 PCUS. To provide for efficient intra-city movement of traffic, the projection of volume provides for grade separated intersections at major crossings, with due consideration of the projected volume of traffic. To provide for smooth movement of traffic along corridors it is envisaged to anticipate construction of grade separated intersections at major locations like Batamaloo junction, Karan Nagar, Moulana Azad Bridge and Kak Sarai. Commercial development presently in vogue in the city should be ordered away from the intersection so that in future, the maintained building lines do not interfere with junction improvement schemes.

Rapid Transit System:

The projected population for Srinagar city by the year 2021 is about 2.3 million. With increasing car-man ratio the personalized mode volume is on the increase exponentially. To reduce dependence on personalized modes it is imperative to provide a mass RAPID-TRANSIT system (MRTS) in future. It is therefore proposed to reserve corridors for a rail based transit system along the North-South direction from Pandach to Nowgam Railway Station via North-South Corridor, Ali Jan Road, Safa Kadal, Karan Nagar Road, Jehangir-Chowk to Tengapora via Batamaloo and along bye-pass to Nowgam. The proposed MRTS shall be rail-based, partly elevated, partly on grades and partly sub-terrain from Karan Nagar-Muincipality road intersection to Batamaloo via Jehangir-Chowk, with dual tracks and terminals. A secondary corridor is also proposed from Saidpora on Ali Jan road to Tengapora on N.H. bye-pass via Achan district-centre, which might be taken up in 13th five year plan period.

There is nowhere in the Master Plan 2000-2021, the mention of the proposal of construction of the bridge over river Jhelum near Convent or that of the fly-over from Jehangir-Chowk to

Rambagh bridge and Natipora crossing. Had the action on the proposals of Master Plan been initiated right in the earnest, there was no need of these slip-shod proposals. How the Govt. violates its own decisions without following the set procedure is not understood and If a law maker himself breaks the laws, how can he enforce the same on others. That is why we are witnessing violations of Master Plan being done by the private parties too, which get regularized later on.

The Master Plan 2000-2021 of Srinagar was approved by the Government vide Cabinet Decision no 11/1 dated 16.01.2003 and was ordered to come into operation from the same very date. It is unfortunate that more than nine years have elapsed; and it is yet stalled at the revision stage.

As regards Convent Bridge, the earlier proposal was got shelved in eighties under stiff opposition by the Institution of Engineers and I remember that a group of engineers led by late Er. Saif-ud-Din Drabu met the then Honb'le CM, and the alignment was later on shifted to Abdullah bridge. Besides the negative points expressed that time being still valid, the present construction is bound to bring the pressure of traffic close to the City-Centre, which is already facing frequent traffic jams. Besides it is going to destroy the recently completed costly beautification of the river banks on either side and that of the historic bund, which has been providing a serene walkway plus the age-old, undisturbed shopping activity for both foreign/local tourists. The raising of approaches on either bank shall be a monstrous structure in the area. Besides the peaceful atmosphere of the prestigious Presentation Convent School is bound to get disturbed affecting adversely the education of our daughters studying there. The alternative site of Zero Bridge or the widening of the existing Abdullah Bridge, if necessary, could be explored which does not violate these considerations.

Since the project details have not been made public for inviting their opinion, no comments can be passed on the skewed alignment of the bridge. The IRC specifications recommend an alignment preferably at right angles to the axis of river. A skew bridge has to resist additional forces due to water pressure and traction. The stresses in a skew slab increase with the angle of skew and the reactions at the supports change with the skew angle. It is found that the specifications for the distribution of wheel loads in right slab and girder bridges are sometimes unsafe and often too conservative.. It has been shown however, that the reaction of an abutment of a 60 degree skew arch uniformly loaded varies from zero to twice the average pressure, which is bound to increase the cost of footings proportionately. A skew alignment has to be resorted to where absolutely essential. In fact even the approaches of the bridge must be in a straight alignment with that of the bridge as per IRC specifications.

Regarding flyover, it is to pass through congested commercial establishments from Jehangir Chowk to Rambagh on both sides of road, which include shops, malls, industrial establishments, schools, religious places, parks, offices and residences rendering these all to be on a lower level with a monstrous structure standing close in their vicinity, also generating noise and dust pollution. Already the Master Plan has spelled out construction of road on Doodganga alignment connecting Batamaloo with Rambagh, which would be a cheaper and viable alternative, besides widening of the existing road from Batamaloo to Tengapora. The estimated cost of 3512 million INR involved in the construction of the flyover could be utilized in a more judicious manner. Besides any amendment in the Master Plan proposals has to be got approved in the cabinet as per the laid procedure.

In the present age of transparency, public opinion needs to be invited on all such projects concerning them, after making all the details public including the environmental impact assessment made by the respective agencies.

Regarding urban infrastructure development I wish our planners pay a flying visit to UAE and watch how the new twin cities of Abu Dhabi and Dubai have come up.

There are 4 to 5-lane roads on either side with a maximum speed of 120 kmph, monitored by CTV cameras with instant dire punishment for violating the traffic rules. There are no street dogs, nor beggars (begging is banned), no solid waste in sight, no dust in the rooms- being all airtight due to AC, no policeman or army man visible, no deafening horns. We find vacuum cleaners cleaning the roads in the night hours, multi-storeyed shopping malls, with multi-storeyed parking lots, round the clock ensured electric supply, water supply and gas connection. Being tax free people from Europe fly in to buy their goods. Last 30-40 years have witnessed conversion of mud huts to high-rise buildings, sky-scrapers and even the world's tallest building Burj Khalifa- 166 Storeys all due to a dedicated ruler late Shaikh Zayed Naihan.

04) WATER AND ENERGY

Water and energy are tightly interlinked and highly interdependent. Choices made in one domain have direct and indirect consequences on the other, positive or negative. The form of energy production being pursued determines the amount of water required to produce that energy. At the same time, the availability and allocation of freshwater resources determine how much (or how little) water can be secured for energy production. Decisions made for water use and management and for energy production can have significant, multifaceted and broad reaching impacts on each other – and these impacts often carry a mix of both positive and negative repercussions.

The challenge today: Extending services to the un-served

Freshwater and energy are crucial for human well-being and sustainable socio-economic development. Their essential roles in achieving progress under every category of development goal are now widely recognized. Major regional and global crises – of climate, poverty, hunger, health and finance – that threaten the livelihood of many, especially the three billion people living on less than US$2.50 per day, are interconnected through water and energy. Worldwide, an estimated 768 million people remain without access to an improved source of water – although by some estimates, the number of people who's right to water is not satisfied could be as high as 3.5 billion – and 2.5 billion remain without access to improved sanitation. More than 1.3 billion people still lack access to electricity, and roughly 2.6 billion use solid fuels (mainly biomass) for cooking. The fact that these figures are often representative of the same people is evidenced by a close association between respiratory diseases caused by indoor air pollution, and diarrhea and related waterborne diseases caused by a lack of safe drinking water and sanitation.

The challenge to come: Meeting growing demands

Demands for freshwater and energy will continue to increase significantly over the coming decades to meet the needs of growing populations and economies, changing lifestyles and evolving consumption patterns, greatly amplifying existing pressures on limited natural resources and on ecosystems. The resulting challenges will be most acute in countries undergoing accelerated transformation and rapid economic growth, or those in which a large segment of the population lacks access to modern services. Global water demand (in terms of water withdrawals) is projected to increase by some 55% by 2050, mainly because of growing demands from manufacturing (400%), thermal electricity generation (140%) and domestic use (130%). As a result, freshwater availability will be increasingly strained

over this time period, and more than 40% of the global population is projected to be living in areas of severe water stress through 2050. There is clear evidence that groundwater supplies are diminishing, with an estimated 20% of the world's aquifers being over-exploited, some critically so. Deterioration of wetlands worldwide is reducing the capacity of ecosystems to purify water. Global energy demand is expected to grow by more than one-third over the period to 2035, with China, India and the Middle Eastern countries accounting for about 60% of the increase. Electricity demand is expected to grow by

approximately 70% by 2035. This growth will be almost entirely in non-Organisation for Economic Co-operation and Development countries, with India and China accounting for more than half that growth.

What rising energy demand means for water
Energy comes in different forms and can be produced in several ways, each having a distinct requirement for – and impact on – water resources. Thus, as a countrys or

Region's energy mix evolves, from fossil fuels to renewable for example, so too do the implications on water and its supporting ecosystem services evolve. Approximately 90%

of global power generation is water intensive. The International Energy Agency estimated global water withdrawals for energy production in 2010 at 583 billion m_3 (representing some 15% of the world's total withdrawals), of which 66 billion m_3 was consumed. By 2035, withdrawals could increase by 20% and consumption by 85%, driven via a shift towards higher efficiency power plants with more advanced cooling systems (that reduce water withdrawals but increase consumption) and increased production of bio-fuel. Local and regional impacts of bio-fuels could be substantial, as their production is among the most water intensive types of fuel production. Despite ongoing progress in the development of renewables, the overall evolution of the global energy mix appears to remain on a relatively fixed path: that of continued reliance on fossil fuels. Oil and gas extraction yields high volumes of 'produced water', which comes out of the well along with the oil and gas. Produced water is usually very difficult and expensive to treat. Unconventional oil and gas production is generally more water intensive than conventional oil and gas production. Thermal power plants are responsible for roughly 80% of global electricity production, and as a sector they are a large user of water. Power plant cooling is responsible for 43% of total freshwater withdrawals in Europe (more than 50% in several countries), nearly 50% in the United States of America, and more than 10% of the national water cap in China.

Similarities, differences and divergences:
Beyond the water–energy nexus
The decisions that determine how water resources are used (or abused) stem from broader policy circles concerned primarily with industrial and economic development, public health, investment and financing, food security and, most relevant to this report, energy security. The challenge for twenty-first century governance is to embrace the multiple aspects, roles and benefits of water, and to place water at the heart of decision-making in all water dependent sectors, including energy. Energy is big business compared to water and can command a great many more resources of all kinds. Market forces have tended to play a much more important role in energy sector development compared with the management of water resources and the improvement of water-related services (water supply and sanitation), which have historically been more of a public health and welfare issue. Water resources have been considered by some to be a *public good* (though the economic definition of 'public good' does not apply to freshwater) – with access to safe water and sanitation recognized as a *human right*. Neither concept ordinarily applies to energy. Reflecting this economic, commercial and social disparity, energy attracts greater political attention than water in most countries. Growing demand for limited water supplies places increasing pressure on water intensive energy producers to seek alternative approaches, especially in areas where energy is competing with other major water users (agriculture, manufacturing, drinking water and sanitation services for cities) and where water uses may be restricted to maintain healthy ecosystems. Uncertainties related to the growth and evolution of global energy production, for example via growth in unconventional sources of gas and oil or in bio-fuels, can pose significant risks to water resources and other users. Policies that benefit one domain can translate to increased risks and detrimental effects in another, but they can also generate co-benefits. The need to manage trade-offs and maximize co-benefits across multiple sectors has become an urgent and a critical issue.

In the context of *thermal power generation*, there is an increasing potential for serious conflict between power, other water users and environmental considerations. Trade-offs can sometimes be reduced by technological advances, but these advances may carry trade-offs of their own. From a water perspective, solar photovoltaic and wind are clearly the most sustainable sources for power generation. However, in most cases, the intermittent service provided by solar photovoltaic and wind needs to be compensated for by other sources of power which, with the exception of geothermal, *do* require water to maintain load balances. Support for the development of renewable energy, which remains far below that for fossil fuels, will need to increase dramatically before it makes a significant change in the global energy mix, and by association, in water demand. Use of geothermal energy for power

generation is underdeveloped and its potential is greatly underappreciated. It is climate independent, produces minimal or near-zero greenhouse gas emissions, does not consume water, and its availability is infinite at human time scales.

Agriculture is currently the largest user of water at the global level, accounting for some 70% of total withdrawals. The food production and supply chain accounts for about one-third of total global energy consumption. The demand for agricultural feeds tocks for bio-fuels is the largest source of new demand for agricultural production in decades, Water planners and decision-makers involved in assessing the water needs of the energy sector require a suitable level of knowledge about electricity generation and fuel extraction technologies and their potential impact on the resource. Energy planners and investors must take into account the complexities of the hydrological cycle and competing water uses when assessing plans and investments.

Thematic challenges and responses

There are many opportunities for the joint development and management of *water and energy infrastructure and technologies* that maximize co-benefits and minimize negative trade-offs. An array of opportunities exists to co-produce energy and water services and to exploit

the benefits of synergies, such as combined power and desalination plants, combined heat and power plants, using alternative water sources for thermal power plant cooling, and even energy recovery from sewerage water. Besides the pursuit of new technical solutions, new political and economic frameworks need to be designed to promote cooperation and integrated planning among sectors. Innovative approaches to spending efficiency, such as cross-sector cooperation to leverage possible synergies, integrated planning for water and energy to decrease costs and ensure sustainability, assessing tradeoffs at the national level, demand-side interventions, and decentralized services, can help overcome the infrastructure financing gap which, although significant for energy, is far greater for water. and was a driving factor behind the 2007–2008 spike in world commodity prices. As biofuels also require water for their processing stages, the water requirements of biofuels produced from irrigated crops can be much larger than for fossil fuels. Energy subsidies allowing farmers to pump aquifers at unsustainable rates of extraction have led to the depletion of groundwater reserves. Applying energy efficiency measures at the farm and at all subsequent stages along the agrifood chain can bring direct savings, through technological and behavioural changes, or indirect savings, through co-benefits derived

from the adoption of agro-ecological farming practices. Knowledge-based precision irrigation can provide flexible, reliable and efficient water application, which can be

complemented by deficit irrigation and wastewater reuse. Many *rapidly growing* cities in developing countries already face problems related to water and energy and have limited capacity to respond. As energy cost is usually the greatest expenditure for water and wastewater utilities, audits to identify and reduce water and energy losses and enhance efficiency can result in substantial energy and financial savings. The future water and energy consumption of a new or an expanding city can be reduced during the early stages of urban planning through the development of compact settlements and investment in systems for integrated urban water management. Such systems and practices include the conservation of water sources, the use of multiple water sources – including rainwater harvesting, stormwater management and wastewater reuse – and the treatment of water to the quality needed for its use rather than treating all water to a potable standard. The chemically bound energy in wastewater can be used for domestic cooking and heating, as fuel for vehicles and power plants, or for operating the treatment plant itself. This biogas replaces fossil fuels, reduces the amount of sludge to be disposed of and achieves financial savings for the plant.

Industry seeks both water and energy efficiency though the two are not always compatible, and though a programme of water and energy efficiency can diverge from industry's primary focus: to secure water and energy at the lowest prices. Individually and together water

and energy efficiency involve varied trade-offs, which frequently involve short-term cost increases against long-term savings, the balance between water and energy use, and a compromise with other factors such as labour, transportation, raw material costs and market location. Large companies and multinationals, particularly in the food and beverage sector, have been engaged for some time in improving water and energy efficiencies. Such

companies see the value of efficiencies in both monetary and societal terms. Small and medium-sized enterprises (with 20 or fewer employees) comprise more than 70% of enterprises in most economies,

and although as a group they have the potential for making a significant impact on water and energy efficiencies, they have fewer resources and are commonly in need of equity capital to do so. The availability of adequate quantities of water, of sufficient quality, depends on *healthy ecosystems* and can be considered an ecosystem service. The maintenance of environmental flows enables this and other ecosystem services that are fundamental to sustainable economic growth and human well-being. Ecosystem services are being compromised worldwide, and energy production is one of the drivers of this process. Natural or green infrastructure can complement, augment or replace the services provided by traditional engineered infrastructure, creating additional benefits in terms of cost-effectiveness, risk management and sustainable development overall.The economic value of ecosystems for downstream water users is formally recognized and monetized in payments for environmental services schemes, in which downstream users provide farmers with payments or green water credits for good management practices that support and regulate ecosystem services, thereby conserving water and
preserving its availability and quality.

Regional priorities
The expansion of hydropower as a key source of renewable energy is a critical issue across nearly all of the world's regions due to concerns of growing conflicts between
various interests over limited water resources. In *Europe and North America*, water scarcity, hydrological variability and the impacts of climate change on water availability and energy production are increasingly recognized as critical – and related – issues. Targets set to increase the share of renewable energies have led to renewed interest in developing pumped storage while part of the region – notably Central Asia and South- Eastern Europe – are still developing new hydropower capacity, not always compatibly with other water uses.
Uncertainties persist over the potential risks to water quality, human health and long-term environmental sustainability from the development of unconventional sources of gas ('fracking') and oil ('tar sands'), both of which require large quantities of water.
be possible and beneficial under certain circumstances, an increased level of collaboration and coordination would create favourable outcomes in nearly all situations.
Effective collaboration does not necessarily require that responsibilities for water and energy are combined into the same institutional portfolio, nor does doing so assure
coherent cooperation. Water and energy practitioners need to engage with and fully understand one another. Both domains have been

traditionally expected to focus on a narrow mandate in meeting their own aims and fulfilling their own targeted responsibilities. There is often little or no incentive to initiate and pursue coordination or integration of policies across sectoral institutions. Policy-makers, planners and practitioners in water and energy need to take steps to identify and overcome the barriers that exist between their domains. The most common responses to the dilemmas, risks and opportunities presented in this fifth edition of the United Nations *World Water Development Report* are related to improving the efficiency and sustainability with which water and energy are used and finding win–win options that create savings of both, which can become mutually

reinforcing (creating synergy). But not every situation offers such opportunities. There are situations in which competition for resources can arise or there is genuine conflict between water and energy aims, meaning some degree of trade-off will be necessary. Dealing with trade-offs may require and benefit from negotiation, especially where international issues are involved. Where competition between different resource domains is likely

to increase, the requirement to make deliberate tradeoffs arises and these trade-offs will need to be managed and contained, preferably through collaboration and in a coordinated manner. To do this, better (and sometimes new) data are required. The incentives to increase efficiency facing the two domains are asymmetrical: energy users have little or

no incentive to conserve water due to zero or low prices, but water users normally do pay for energy, even though prices may be subsidized. Water and energy prices are strongly affected by political decisions and subsidies that support major sectors such as agriculture and industry, and these subsidies often distort the true economic relationship between water and energy. Particularly for water, price is rarely a true reflection of cost – it is often even less than the cost of supply. With its demand for energy increasing exponentially, the *Asia-Pacific* region faces major supply challenges. Coal, the most prevalent energy product within the region, will continue to be the main source of energy, despite serious concerns about water quality degradation as an effect of coal mining and the water quantity required for cooling thermal power plants. The potential for Asia to develop into a significant market for and exporter of biofuels is being increasingly recognized, and there is a hope that it will provide new employment opportunities in several developing nations.

In the *Arab region*, the low to middle income countries are struggling to meet growing demands for water and energy services. Limited understanding of the interdependencies affecting the management of water and energy resources has stymied coordination between water and energy policy-makers, and limited coordination between the water, energy, electricity and agriculture sectors has led to conflicting policies and development objectives. Solardriven desalination and energy recovery from wastewater are two promising technologies particularly well suited to the region.

In *Latin America and the Caribbean*, there is an increasing interest in biofuels and in how more water efficient (and more energy intensive) irrigation methods and electricity subsidies to farmers impact on aquifer sustainability. The vast majority of water utilities in the region are struggling to attain self-financing and, as energy is often the greatest component (30–40%) of operational costs, increasing energy costs have direct implications for service affordability and for sector financing.

The majority of the rural population in *sub-Saharan Africa* relies on traditional energy supplies, mainly unprocessed biomass, the burning of which causes significant pollution

and health concerns. It is the only region in the world where the absolute number of people without access to electricity is increasing. As Africa has not yet tapped in

to its rich potential for hydropower development to a substantial degree, it has the opportunity of learning from the positive as well as the negative aspects of hydropower

implementation practices that other nations have undergone.

Enabling environments

Enabling environments

Recognition of the interconnectedness between water and energy has led some observers to call for a greater level of integration of the two domains. Although this may

A coherent policy – which is to say an adequate public response to the interconnectedness of the water, energy and related domains – requires a hierarchy of actions.

These include:

· Developing coherent national policies affecting the different domains

· Creating legal and institutional frameworks to promote this coherence

· Ensuring reliable data and statistics to make and monitor decisions

· Encouraging awareness through education, training and public information media

· Supporting innovation and research into technological development

· Ensuring availability of finance

· Allowing markets and businesses to develop

Together these actions make up the *enabling environment* necessary to bring about the changes needed for the sustainable and mutually compatible development of water and energy. The international community can bring actors together and catalyse support for national, subnational and local governments as well as utility providers, who have a major role in how the water–energy nexus plays out at the national and local levels. The different political economies of water and energy should be recognized, as these affect the scope, speed and direction of change in each domain. While energy generally carries great political clout, water most often does not. Partly as a result, there is a marked difference in the pace of change in the domains; a pace which is driven also by the evolution of markets and technologies. Unless those responsible for water step up their own governance reform efforts, the pressures emanating from developments in the energy sphere will become increasingly restrictive and make the tasks facing water planners, and the objective of a secure water future, much more difficult to achieve. And failures in water can lead directly to failures in energy and other sectors critical for development.

Water and energy are closely interconnected and highly interdependent. Choices made and actions taken in one domain can greatly affect the other, positively or negatively. Trade-offs need to be managed to limit negative impacts and foster opportunities for synergy. Water and energy have crucial impacts on poverty alleviation both *directly*, as a number of the Millennium Development Goals (MDGs) depend on major improvements in access to water, sanitation, power and energy sources, and *indirectly*, as water and energy can be binding constraints on economic growth – the ultimate hope for widespread poverty reduction. In view of the post-2015 Sustainable Development Goals, likely to include increased access to water and energy services, this fifth edition of the United Nations *World Water Development Report* seeks to inform decision makers (inside and outside the water and energy domains), stakeholders and practitioners about the inter linkages, potential synergies and trade-offs, and to highlight the need for appropriate responses and regulatory frameworks that account for both water and energy priorities. The Report provides a comprehensive overview of major and emerging trends from around the world, with examples of how some of the trend-related challenges have been addressed, their implications for policy-makers, and further actions that can be taken by stakeholders and the international community.

Part 1 of the Report:'Status, trends and challenges, explores current and future challenges of sustainable development in the context of ever-increasing demands for water and energy. Chapter 1 describes many of the complex inter linkages between the water and energy domains from varied perspectives, highlighting their interdependencies and differences, as well as their relationships to other developmental sectors. Chapter 2

focuses on water, examining current and future demand and the pressures that drive demand as well as the energy requirements for the provision of water services. The chapter also provides a snapshot of the state of water resources using the latest information available. Chapter 3 examines sources of energy, both renewable and nonrenewable, and existing means of power generation in terms of their current and future contribution to the global energy mix and their impact on water. Chapter 4 focuses on data and knowledge issues directly related to the water–energy nexus, highlighting the need to generate

and harmonize data concerning the supply and use of water and energy production. In the future, growing demands on water resources resulting from population growth, economic development and urban expansion will create additional pressure on water intensive energy production. Climate change will add to the pressure. Droughts, heatwaves and local water scarcities of the past decade have interrupted electricity generation, with serious economic consequences. At the same time, limitations on energy availability have constrained the delivery of water services. Growing demand for finite water resources is also leading to increased competition between the energy sector and other water-using sectors of the economy, principally agriculture and industry. A very important aspect of

the burgeoning global demand for water services is the resulting pressure on water resources and the degradation of freshwater ecosystems.

Part 2, the 'Thematic focus', narrows the examination of water and energy into five specific themes. Chapter 5 looks into the economic aspects of water and energy infrastructure in developed and developing countries, highlighting some opportunities for synergies in infrastructure development, operation and maintenance. The challenges and response options faced by food and agriculture, including biofuels, in relation to water and energy are presented in Chapter 6. Chapter 7 focuses on the particular difficulties facing the large and rapidly expanding cities in developing countries. Chapter 8 describes the role of industry as both a user of water and energy but also a potential leader in the development of innovative approaches to efficiency. Part 2 concludes with Chapter 9, which argues that ecosystems are the foundation of the water–energy nexus and that an ecosystem approach is vital for green growth.

'Regional aspects' of water and energy are provided in Part 3. Issues of concern for Europe and North America, from expanding hydropower and its related conflicts to the development of unconventional sources of oil and gas, are presented in Chapter 10. Chapter 11 describes how increasing reliance on coal and the development of biofuel in the Asia-Pacific region will impact on water resources and other users. The need to increase knowledge and raise awareness for policy coherence and the potential of certain water supply and treatment technologies in the Arab region are addressed in Chapter 12. Chapter 13 examines hydropower development and the energy requirements for water services in Latin America and the Caribbean. Hydropower is also the focus of Chapter 14, which highlights the urgent need to increase access to electricity in sub-Saharan Africa where the undeveloped potential for hydropower is the greatest of any region. Where competition between different resource domains is likely to increase, the requirement to make deliberate trade-offs arises, and these trade-offs will need to be managed and contained, preferably through collaboration and in a coordinated manner. There are, fortunately, already good examples of policies and actions that benefit both water and energy domains, such as win–win projects and optimized trade-offs.

Part 4, 'Responses: Fostering synergies and managing trade-offs', describes how policy-makers, decision-makers and practitioners can respond to the dilemmas, risks and opportunities presented in the first three parts of the Report. Chapter 15 proposes a hierarchy of actions that together make up the enabling environment necessary to bring about the changes needed for the sustainable and mutually compatible development of water and energy. These actions include overcoming the barriers that exist between the two domains, using economic instruments appropriately, and optimizing the role of the United Nations system and the international community. The Report concludes with Chapter 16, in which the interplay of water and energy, and the scope for fostering synergies and managing trade-offs between them, is illustrated in the contexts of agriculture, industry, cities, ecosystems and power generation.

05) Small Islands & Climate Change

The UN General Assembly declared 2014 as the International Year of Small Island Developing States (SIDS). World Environment Day (WED) 2014 will be celebrated under the theme of SIDS, with the goal of raising awareness of their unique development challenges and successes regarding a range of environmental problems, including climate change, waste management, unsustainable consumption, degradation of natural resources, and extreme natural disasters.

Climate change is a major challenge for SIDs, as global warming is causing ocean level to rise. Due to their small size and isolation, SIDs are more vulnerable to natural & environmental disasters, climate change & sea level rise. However some of these islands have also been successful in overcoming their environmental problems. From Palau to Puerto Rico, the stories of resiliency and innovation abound. For instance, Tokelau recently began producing 100% of its energy from solar sources. In Fiji, lacking the resources to make new drainage systems and seawalls, local residents are restoring mangroves and coral reefs to help prevent flooding and erosion. These stories and solutions can be applied to environmental concerns all over the world. The problems that they face are: climate change, waste management, unsustainable consumption, degradation of natural resources, extreme natural disasters in most of over population and continuing industrialization.

We in J&K State have witnessed an elongated winter this year with frequent rains and snowfall in hilly areas even in the month of May, due to climate change. Even in the month of June we still feel comparatively lower temperatures. Besides the other threats of pollution of our water bodies, air & noise pollution,

inadequate solid & liquid waste management, forest denudation
etc. persist despite it being branded as a heaven on earth.

The Ocean edge:

Over the long periods of time, the level of the sea changes either
drowning what has been land or exposing what has been sea bed.
The tales of catastrophic floods told in the folklore and legends
of many cultures are corroborated by scientific evidence. The sea
level of today is 500 ft. (150m) higher than it was 25000 years
ago, yet in the context of global history, this is one episode in the
ebb & flow of the shore, as the land emerges or is submerged,
depending on the relative level of the land and sea.

Worldwide change of the sea level however may be the result of
melting in the polar ice caps or changes in the capacity of ocean
basins as sediment from the is disgorged by rivers into the sea.

Waves at work:

The turbulence of the wind stirs the surface of the sea, waves are
the result. When waves that have traveled hundreds of miles
across the ocean encounter the coast, the steady swell of their
journey is broken in a matter of yards. The impact of this
collision is the single greatest factor in shaping our shores, as the
waves lap gently at the beaches or pound the cliffs with a force
of up to 30 tons per sq. meter. Stones and pebbles carried in the
water are the wave's cutting edges reputed to have shattered
lighthouse windows 100 m. up, sawing and grinding at the cliff
faces & gradually pushing back the line of the coast.

Stop killing our oceans:

In her book "The Sea Around us" Rachel Carson saw the oceans
as one last haven safe for ever. How it could be otherwise, when
the oceans are so vast that continents are mere islands in their

midst, so deep that a Mount Everest could be lost beneath their surface? How does one pollute a volume of almost 320 million cubic miles? How poison an environment so rich that it harbors 200,000 species of life?

Even though the oceans blanket three fourths of earth, their productivity is limited mostly to the narrow bands of undersea land existing from coastlines which comprise the continental shelves. 80% of the world's salt water – fish catches taken from these shallow coastal waters, which make up only a tiny fraction of the total sea. In addition almost 70% of all usable fish & shell fish spend a crucial part of their lives in the estuaries- the coastal bays, tidelands & river mouths that are 20 times more fertile than the open sea, seven times more fertile than a wheat field. Cut the chain of life in these areas, destroy the myriad bottom organisms, pollute the continental shelf waters & you will also eliminate the vital ocean fisheries.

Already pollution or overfishing & sometimes both have gouged fisheries around the world. Meanwhile in a headlong rush to create more land , vital coastal tidelands are being filled for highways, industry, bridges and water front homes. At the same time the remaining estuaries are fed billions of gallons of sewage & industial waste every day. These poison fish, choke our oyster & render the bays & tidelands unfit for anything.

Pressure also builds on the oceans beyond, for instance in 1968 some 48 million tons of solid wastes were carried out by barges and ships& dumped in ocean waters of the USA. These wastes included garbage, waste oil, dredging spoils, industrial acids, caustics, cleaners & sludges, airplane parts, junked automobiles & spoiled food. During the two papyrus boat trips across the Atlantic, author explorer Thor Heyerdahl sighted plastic bottles, squeeze tubes, oil & other trash that had somehow been swept by the currents to mid-ocean. On some days the crew hesitated to wash because of the amount of pollution. Ugly brown raw sewage is piped from Miami beach Florida that sprawls over blue green waves. Fishermen, divers & others report similar situations all along US coast lines & many other parts of the world.

Steps needed:

- Stop dumping of wastes in to the sea, the big & small lakes & coastal areas, our rivers & bays except for treated liquid wastes equal to natural quality of the ocean waters. Instead recycle wastes back in to economy.
- Set tough controls before undertaking new ocean activities such as buildings, off shore jet ports & drilling off shore oil wells in new areas.
- Halt the reckless dredging & filling of priceless tidelands & carving of ocean front in the name of progress.

With the present trend the marine biologists predict that it will put an end to significant life in the sea in 50 years or less. This would be a catastrophe posing grave consequences to a world dependent on these vital resources for food, raw materials, recreation & in the near future, probably living space.

The **largest island** in the world is Greenland. Australia is considered a continent because it has unique plant and animal life. Antarctica also is a continent – larger than Europe and Australia. Greenland, although quite big, shares the habitat features of Northern America.

The **smallest island in the world** – according to the Guinness Book of Records – is Bishop Rock. It lies at the most south-westerly part of the United Kingdom. It is one of 1040 islands around Britain and only has a lighthouse on it. In 1861, the

British government set out the parameters for classifying an island. It was decided that if it was inhabited, the size was immaterial. However, if it was uninhabited, it had to be "the summer's pasturage of at least one sheep" – which is about two acres.

A lot of standing room – not much else. This is Bishop Rock, the world's smallest island.

Going by the above parameters, most of the 179 584 "islands" around Finland and the almost 200 000 around Canada would not match Indonesia as the country with the most islands. In fact, Indonesia consists only of islands – 13 667 of them, 6000 of which are inhabited.

The **remotest uninhabited island** is Bouvet Island in the South Atlantic. The remotest inhabited island in the world is Tristan da Cunha. It is in the South Atlantic, 2575 km (1600 miles) south of St Helena, which is an island a few hundred kilometres (miles) off the coast of South Africa. Tristan da Cunha has no TV but it has one radio station. The population totals 242 and they only have 7 surnames (last names) between them, so they are all related. Tristan da Cunha does have a capital, called Edinburgh of the South Seas.

The **smallest independent island country** is the Pacific island of Nauru. It measures 21,28 sq km (8.2 sq mi). (Only the Vatican City and Monaco are smaller countries.)

Of the 6 billion+ people in the world, one out of ten lives on an island (600 million). Which is not so hard to imagine when you consider that more than 240 million people live in Indonesia alone – and some 61 million live in Britain, the only island connected to a continent through the chunnel (tunnel under the sea).

"In the 13th Century, people convicted of a serious crime were taken out to Bishop Rock and left there with bread and water to be ravaged by the sea. Or so history suggests. Today the rock is better known as the point where record-breaking attempts to cross the Atlantic are started and finished." –

During British rule freedom fighters of India were punished by sending them to the prison of Andaman and Nicobar islands called "Kala Pani" i.e."The black water" and were inhabited by cannibals.

Iceland, ironically, is quite green. Over one third of Iceland is volcanically active and loaded with lava fields. Iceland is far enough north to be entirely covered by ice, like Greenland to the west of it, but the magma below the surface heats the rock above, keeping it "green."

Greenland is the largest island in the world- other than its name suggests, is covered in ice- although some of it is melting.

The world's largest islands by area size:

1. Greenland, Denmark – 836 330.48 sq miles (2,166,086 sq km) – coastline: 27,394 miles (44,087 km).
2. New Guinea (Papua New Guinea and West New Guinea) – 303,381 sq miles (785,753 sq km).
3. Borneo – 288,869 sq miles (748,168 sq km).
4. Madagascar – 226,658 sq miles (587,041 sq km) – coastline: 3,000 miles (4,828 km).
5. Baffin Island (Qikiqtaaluk), Canada – 195,928 sq miles (507,451 sq km).
6. Sumatra, Indonesia – 171,069 sq miles (443,066 sq km).
7. United Kingdom – 94,058 sq miles (243,610 sq km) – coastline: 7,723 miles (12,429 km).
8. Honshu, Japan – 87,182 sq miles (225,800 sq km).

9. Victoria Island, Canada – 83,89 sq miles (217,291 sq km).
10. Ellesmere Island, Canada – 75,767 sq miles (196,236 sq km

TOP 10 smallest countries in the world

The number of countries is increasing every year because more and more of them become independent and so we get many new countries. The number of independent countries considers being currently 196. 193 countries are those that are officially members of the United Nations.

Countries have different sizes, but there are quite a few very large countries that are not densely populated. Some countries are very influential, while others are popular for tourism and natural attractions. Today, we shall look at in more detail micro-states. Those are the smallest countries in the world or they are with very small population.

1. Vatican city
The absolute winner among micro-countries is the Vatican.
Vatican is the smallest independent country in the world.
It has an area 0.44 km^2.
It is the smallest by the number of the number of inhabitants too.
The population is just over 800.
Vatican City is surrounded by city of Rome.
Vatican is the most famous for that it is the home of the Pope.
Despite its small size the Vatican extremely popular tourist destination visited by millions of tourists each year.
The Vatican has its own telephone system, mail boxes, gardens, radio station, banking system and pharmacy
The Vatican has Swiss Guard responsible for the personal safety of the pope since 1506th.
Almost all supplies, including food, water, electricity and gas must be imported.
There is no income tax and no restrictions on the import or export of resources.
Their income is from voluntary contributions from more than one billion Roman Catholics worldwide.
2. Monaco
The second smallest country is Monaco.

It has an area of 1.98 km^2 and it is like the Vatican despite the small size very a popular tourist destination.

The country has a mild Mediterranean climate with an average annual temperature of 16 ° C and about 60 days of rain.

Monthly average temperatures are in range from 10 °C in January to 24 °C in August.

It is most famous for Formula 1 races which have been ongoing since 1929.

Monaco is the most populous country in the world.

It is home to many rich people who have been moved there because the country has no income tax.

Most of the population of Monaco is made up of French citizens (almost half); small but significant number of Italians, Swiss and Belgians.

Most people are Roman Catholics.

The official language is French.

3. Nauru

The third smallest state is probably less known.

Nauru is a small island country that lies in the southern part of the Pacific Ocean.

Its specialty is definitely that it is the only country in the world that does not have an official capital city.

Country's total area is only 21 km^2 and consists of coral islands.

People there don't have to pay taxes which had a positive impact on the standard of living of the population.

After Vatican City it is the second least-populated country.

They have only 9,378 residents.

The electricity is free.

The island was discovered by English captain John Fearne in the year 1789.

4. Tuvalu

Tuvalu is also an island country.

It has an area of only 26 km^2.

It is located halfway between Hawaii and Australia.

The biggest problem of Tuvalu as a country that sea levels are rising because the islands lie on average five meters above sea level and therefore they are threatened that they will soon sink.

The main income of the distant country is a foreign financial assistance.

Some of the income they have earned by selling telephone code 900 and an internet domain tv.

5. San Marino

The following micro-state is San Marino.
San Marino is surrounded by Italy.
It has an area of 61 km^2.
It is the oldest country in the world.
It is also the oldest Republic in the world.

6. Liechtenstein

Liechtenstein is one of the micro-states located in Europe.
It is an alpine country between Switzerland and Austria with an area of 162 61 km^2.
Liechtenstein is a principality.
It is known around the world mainly for its banking and tax benefits so that the smallest and richest country.

7. Marshall Islands

Marshall Islands are a set of 34 islands in the central Pacific Ocean.
They are mainly interesting because they don't have strict administrative regulations.
Marshall Islands are known mainly due to nuclear testing on the islands by the United States of America.
The islands are extremely beautiful by the pictures and have crystal blue bays and sandy beaches.

8. Saint Kitts and Nevis

Saint Kitts and Nevis are also islands.

They have an area of 261 km^{2}and lies at the crossroads of the Caribbean Sea and the Atlantic Ocean.

Saint Kitts are also known more formally as Saint Christopher Island.

For a long time the locals most of the income produced by the sugar but today is an important sector of income also agriculture and banking.

9. The Maldives

The Maldives is one of the dream destinations.

Maldives are spread over 300 km^2 in the Indian Ocean.

All together the Maldives consists of more than 2,000 atolls.

Only 218 islands are inhabited.

10. The Republic of Seychelles

The Republic of Seychelles is a set of about 115 islands in the Indian Ocean.

They have total area of 450 km^2.

Seychelles is one of the most popular destinations for honeymoon and they are considered as a dream destination - long, sandy beaches, good food and clean sea.

Here I would like to reproduce the extracts of the report of Mr. Ronny Jumeau Seychelles Ambassador for Climate Change and SIDS issues on Expert Group Meeting on Oceans, Seas and Sustainable Development: Implementation and follow-up to Rio+20 at United Nations Headquarters 18-19 April 2013 and here I quote:

"While oceans play a key role in everyone's lives, no one is more dependent on them than the small, vulnerable and isolated island developing states surrounded by the seas. Oceans are now firmly established on the global agenda after taking center stage at Rio+20 last year. However, the SIDS' unique dependence on the marine environment means the oceans have commanded center stage in our development since humankind has been on the islands. We are the ocean people, so to speak: we live off and by the oceans and to varying degrees on and for them as well. The oceans define who we are and the coastal and marine environment is an integral part of our island lifestyle. Our islands may be small in land area, but we morph into large ocean states when our exclusive economic zones are factored in. Tuvalu's EEZ for example, is 27,000 times the size of its land. The Republic of Kiribati, the largest small island developing state in terms of ocean territory, has the 13th largest exclusive economic zone on Earth. SIDS are the custodians of no fewer than 15, or 30 per cent, of the 50 largest EEZs.

Dependence on oceans

In the case of many islands, Seychelles, our no 1 pillar of the economy is marine-based tourism. It provides 26 per cent of GDP, 30 per cent of jobs and 70 per cent of foreign exchange earnings in a country where more than 80 per cent of what we consume is imported, mostly by sea. Fisheries, our second biggest industry, add another eight per cent to our GDP. Such a heavy dependence on oceans is repeated across the SIDS. Oceans are central to our sustainable development, to poverty reduction and achieving the Millennium Development Goals, and to our post-2015 development agenda. And yet, despite our best efforts to help ourselves, the lack of technical, institutional, technological and financial support means we are still to benefit to the fullest from our marine resources. Where we do benefit, it is not necessarily in the most sustainable manner.

Rio+20 :

It is no surprise therefore, that the small island developing states formed the loudest cheering section when the oceans won big at Rio+20.There definitely needs to be an international instrument regulating the conservation and sustainable use of marine biodiversity in areas beyond national jurisdiction. SIDS therefore welcome Rio+20's call for a United Nations General Assembly decision to develop such an instrument under the Convention on the Law of the Sea by next year.

Marine Pollution

Nowhere are the effects of marine pollution more deeply felt and damaging than in the small island countries entirely surrounded by the ocean. This is especially so in SIDS like mine whose economies are heavily dependent on the state and indeed the attractiveness of our beaches, coastal waters, coral reefs and fisheries.

Sea Level Rise

The most serious long-term threat to SIDS is of course sea level rise which threatens to cover many of us with the oceans, thus turning the blue planet even bluer…that is if we are not swept away first by coastal erosion which will be made worse by the slow collapse of dying reefs.

Ocean Acidification

Ocean acidification is the single greatest threat to coral reefs which provide SIDS with food and income. They also protect us from the ocean waves and tidal currents which, as extreme weather events such as storm surges and abnormally high tides intensify, threaten to sweep away some islands before they are covered by sea level rise.

Coral Reefs

Rio+20 drew attention to the important economic, social and environmental contribution of coral reefs, especially to islands and other coastal states, and the high vulnerability of coral reefs and mangroves to such impacts as climate change, ocean acidification, overfishing, destructive fishing practices and pollution, among others. Indeed, the growing pressures on coral reefs may cause them to be the first marine ecosystem to collapse.

Marine Protected Areas

SIDS thus see conservation measures such as marine protected areas not just as a way to protect our ocean biodiversity and resources, but also as a tool for sustainable development, because for us marine biodiversity has significant socio-economic and cultural value.

Fisheries

Finally on fisheries, I would like to return to the Rio+20 outcome document The Future We Want, specifically paragraph 168. In it we commit to intensify efforts and take measures to meet the Johannesburg Plan of Implementation's 2015 target to maintain or restore stocks to levels that can produce maximum sustainable yield in the shortest time feasible. Once again the effect of illegal, unreported and unregulated or IUU fishing is most felt in those countries that depend most heavily on fisheries like the small island developing states. We place strong emphasis on paragraph 174 of The Future We Want. This urges that by next year there be strategies to further help developing states, especially the least developed and SIDS, develop their national capacity to conserve, sustainably manage and realize the benefits of sustainable fisheries, including through improved market access for their fish products. I cannot over-emphasize the importance of this to small island developing states. In the Pacific

SIDS, for example, the tuna fishery alone contributes more than 10 per cent of GDP and in some islands more than 50 per cent of exports. It is estimated that fish contributes at least 50 per cent of total animal protein intake in some SIDS. There certainly is no lack of international instruments in fisheries: they cover straddling and highly migratory fish stocks, responsible fisheries and IUU fishing. What has been lacking is the political will to effectively implement and enforce them. As I showed in the examples I gave earlier, SIDS certainly do not lack political will or innovative thinking: what we lack is capacity – technical, institutional, technological and financial" -en quote.

About oceans, Allah mentions in Quran:

مرج البحرين يلتقيان ه بينهما برزخ لا يبغيان فبآى آلأى ربكما تكَذ بان ه يخرج منهما اللولو والمرجان ه فبـاى آلاى ُ ربكما تكذبان ه

He has let free the two bodies of flowing water (seas), meeting together. Between them is a Barrier, which they do not transgress. Then which of the favors of your Lord will you deny? Out of them come Pearls and Coral. Then which of the favors of your Lord will you deny? (Surah 55, Ayat 19-23)

06) 2014 KASHMIR FLOODS: NEED FOR INITIATING CORRECTIVES

(A) 'Floods can be prevented if the carrying capacity of the river Jhelum and the spill channel is increased by dredging in a time-bound schedule.

01) *Cause of recent floods:* Rarely Septembers have witnessed such a heavy downpour. It appears that the much talked about global climate change has played a major part in this unusual/unprecedented behavior of weather. Met Dept. reported a record rainfall of 400 mm, 225 mm on a single day combined with cloudbursts in the upper catchments of South Kashmir raising the flood level at Sangam to 16 ft. higher than the danger mark of 21ft., beating all previous records. Consequently the gauge at Rammunshi Bagh touched 11 ft. higher than the danger mark of 18 ft. This caused overflow of the flood water on the river banks and subsequent breaches. Since Rajbagh is located below water level of abutting river Jhelum and flood spill channel, it is natural for water to rise from the domestic soakage pits even in normal days in order to maintain its water table.

However, it is but natural for water to overflow its banks in the event of rainfall in its upper catchment and spill into flood plains which are basically its right of way. Extensive and often unplanned use of flood plains, disregarding the basic fact that it is a part and parcel of the river, leads to flood damage. Thus the uncontrolled and indiscriminate development of flood plains due to pressure of population can be considered as one of the main factors responsible for the ever increasing flood damage in spite of the substantial investment in the flood-sector during the last six decades.

Arresting the gushing flow from the tributaries by way of construction of check dams and creating storage within the provisions of Indus Water Treaty could check the instant swelling of the river discharge.

Conversion of Bemina wetland to housing colony by resorting to 7 ft. filling was strongly resented by the then Chief Engineer I&FC, but his opinion was overruled by the then CTP and others. The cross section of river Jhelum has got reduced by recent beautification which was debated by the Institution of Engineers India J&K State Centre.

Flood spill channel and Jhelum embankments have been encroached upon by the public as well as the Government. Public is of the opinion that there has been delay in breaching the bund at Kandizal to relieve the intensity of flow of flood waters. The dredging of the river and spill channel to increase their carrying capacity can be a less costly solution.

02) Lessons from the past experience:

Based on the past experience, people would state that a major flood is expected after every half a century. Lawrence has recorded in his "Valley of Kashmir" the worst flood of 1903. Next worst flood was witnessed in 1959 i.e after 56 years. So after 55 years, the clouds of the major flood were looming large in the air according to this theory. Much was being talked, about disaster management and its preparedness after the last earthquake in Uri and Kupwara sectors and some people were reported to have been trained at district levels too, but none was seen on the trouble spots except the local youth who have shown an exemplary bravery to save the people in distress. Though Ministry of Home Affairs had initiated NDRM programme in all the flood prone states, providing assistance to draw disaster management plans at the State, District, Block and village levels, yet it yielded no results. Further the loss of total communication added to the misery of people.

About flood duty preparedness, it was the usual practice that a flood control committee comprising of the concerned heads of departments would be nominated every year. Besides flood duty charts would be circulated every year beforehand which would assign engineers of all departments and their other staff their respective beats. Every year there would be a day fixed for flood rehearsal, when river banks would be examined and necessary strengthening of the weak spots recommended besides arrangement of trucks, storage of empty bags, boats etc. ensured. Also the location of flood beats would be conveyed to the staff according to the flood duty chart. On declaration of floods, responsible officers would man the control room, monitoring the latest situation including on-spot inspections round the clock and passing the directions regarding deployment of labour, dispatch of empty bags, rescue boats, trucks etc. The activities were monitored at the highest level and the CM was kept informed about the latest situation. Important decisions to save the population would be taken at highest level. Messages would be conveyed on phones, wireless followed by written messages/orders for record and reference. The flood duty was not called off until the floods would recede completely and the affected people rescued and provided with the necessary relief.

03) Fulfilling housing needs of the people:

The ideal solution would have been to plan a new city on higher contours along foot-hills from Ganderbal to Harwan, Pandrethan, Khunmoh to Rajpora and on Karewas along south side ensuring safety from river floods, proper drainage, sunlight and fresh mountain breeze but again that may prove to be a herculean task.

Given the hunger for land the possibility of vertical expansion as already advocated for the city should be considered seriously to tackle the future housing needs in view of multiplying population, besides relocating the families of flood hit/prone areas. The basement floors could be left for parking of vehicles.

04) *Possibility of saving the city in recent floods*:

People are of the opinion that the bank at Kandizal was not breached in time to divert part of flood waters as per the past practice. Besides, the weak bunds of the river were not strengthened well in advance in spite of threat of major floods. A road has come up over pipes right on the bed of flood spill channel that has caused an obstruction to the flood waters. In the past Dal Lake would also serve as a receptor of some waters when its level was maintained at a lower level than the present one. People are also doubtful if the gates at Nalla Amir Khan, Chattabal weir, Dood ganga diversion were opened well in time.

05) Measures required for rebuilding houses:

Now that the centuries' worst flood has receded, people shall be rebuilding their damaged houses in their respective sites. Some important points need to be considered to prevent recurrence of such an eventuality in future:

People in present distress cannot wait for the long term measures of flood protection and they will soon resort to the reconstruction of their collapsed/damaged houses. In view of approaching winter, houses of prefabricated members could be erected urgently at suitable locations or rented accommodation could be provided to them till their permanent structures come up in the next summer season.

The Govt./SMC must come up with the new building norms, that are required in the flood prone areas to ensure the safety of lives of the inhabitants. In framed structures, basement floors could be raised on RCC columns with ceiling level higher than the HFL, leaving the space for car parking etc. In fact such norms already provided for shopping complexes have been violated causing parking problems in the city. SMC must implement these norms strictly.

Safe foundations as per bearing capacity tests need to be designed as per BIS specifications with the approval/check of SMC. In fact I found a ten-storey structure was stalled by Muncipal Authorities at plinth level in Abu Dhabi, the reason being provision of lesser steel than the approved/designed one.

The partially damaged structures of the flooded area need to be inspected by an expert team to suggest measures for their retrofitting. Dewatering stations need to be lifted higher than the HFL of 100 year flood for making these functional during crisis.

The plinths of all Govt. buildings/hospitals/establishments in the flood prone areas need to be raised higher than the HFL. The basement floors could be used for car parking etc.

In masonry structures, damp proof course as per the prescribed specifications with horizontal/vertical mastic asphalt layer could prevent dampness and at least 9 inches RCC beam at DPC level could hold a two feet gap of sinking foundation.

Flood plain zoning is useful in reducing the damage caused by drainage congestion particularly in urban area where on grounds of economy and other considerations urban drainage may not be designed for the worst possible conditions and presupposes some damage during storms

whose magnitude exceeds that for which the drainage system is designed.
The steps involved in implementation of flood plain zoning measures could be:
Demarcation of areas liable to floods. Preparation of detailed contour plans of such areas to a large scale (preferably 1:5000) showing contours at interval of 0'3 to 0.5 meters.
Fixation of reference river gauges and determination of areas likely to be inundated for different water levels and magnitudes of floods. Demarcation of areas liable to flooding by floods of different frequencies should be done, like once in two years, five, ten, twenty, fifty and hundred years and similarly areas likely to be affected on account of accumulated rainfall, like in 5, 10, 25, and 50 years

06) Steps needed for minimizing damage:

In the existing developed areas possibilities of protecting, relocation and exchanging the sites of vital installations like electricity substations/powerhouses, telephone exchanges etc. should be seriously examined so that these are always safe from possible flood damage. Similarly the pump stations of tube wells for drinking water supply should be raised above the HFL corresponding to a 100 year flood.

Similarly possibility of removing buildings/structures obstructing existing natural drainage should be seriously considered. In any case unplanned growth shall be restricted so that no constructions obstructing natural drainage resulting in increased flood is allowed. In future the following regulations may be stipulated:
Plinth levels of all buildings should be nearly 0.75 to one meter above the drainage/submersion levels.

In the areas liable to floods, for all the buildings a stairway should be provided to the roofs/attic floors so that temporary shelter can be taken there. The roof levels of the single storey buildings and the first floor level in double storey buildings should be above flood level of 1 to 100 frequency so that the human lives and the movable property can take temporary shelter there when necessary during the floods.

In the past Central Water Commission (CWC) prepared guidelines in 1873-74 for flood plain zoning which were approved by Central Flood Control Board. CWC also prepared a model draft and circulated it in the Ministry of Irrigation in 1975, to all the states for enacting legislature. However the response from states, except Manipur, has not been encouraging. Manipur enacted a legislation in 1978 which came into force in 1985.

Flood proofing measures help greatly in the mitigation of distress and provide immediate relief to the population in flood prone areas. It is essentially a combination of structural change and emergency action, not involving any evacuation. The technique adopted consists of providing raised platforms for flood shelter for men and cattle and raising the public utility installation above flood levels.

In case of urban areas, certain measures that can be put into action as soon as a flood warning is received involve: Installation of removable covers such as steel or aluminium bulk heads over doors and windows or other openings keeping stone counters on wheels, closing of sewer well, anchoring machinery, covering machinery with plastic sheet, seepage control etc. Flood proofing also tends to encourage persistent human occupation of flood plains.

07) Present response of the people at the helm:

The abrupt and unexpected flash flood has caught us all unawares. Once the flood alarm was sounded, all concerned were alerted and took their respective positions but it was an eleven ft. high wall of water above danger mark that swept away the flood control room itself and disruption of the communication added to the misery. It has given sleepless nights to one and all. However a quick response would have been relocation of the control room at a higher/safer place and provision of an alternative wireless communication system which seems to have been abandoned now. The damage could have been minimized by timely breaching of Kandizal bund that would have reduced the intensity of flood flow. Besides sounding of alarm on Radio, TV, loudspeakers in mosques could have been helpful. The concerned department must have sufficient boats, even motor boats, reserved for such an eventuality. The mechanized water transport would have been of great help had it been executed as per the project.

08) Immediate *dredging and desilting of river bed and FS channel is recommended.:*

There is no guarantee that floods of even worse magnitude may not visit the valley again. The global climate change is one of the major factors.

Government seems to be contemplating to provide a parallel spill channel on the upstream side of the existing one, but that may involve a huge investment besides being a long drawn affair of time consuming land compensation etc. Besides this, the channel may also get defunct with the passage of time like the present one for want of its maintenance. My viewpoint has been seconded in this piece by Er. Joseph Thomas of Bangalore

http://amolak.in/web/flood-prevention-in-the-vale-of-kashmir-by-joseph-thomas/#comments.

He has commented: "I am surprised to see newspaper reports that the state government wants to divert the Jhelum river around Srinagar. It is feasible but at prohibitive cost. Firstly, a lot of land is required. Secondly, a large number of bridges will have to be built across the new river channel. Thirdly, when the new river channel gets silted up, embankments will be built on either side. Eventually, the river will again flow above street level. We will be back to square one."

It is far better to dredge and de-silt the existing river and flood channel. No additional land is required and new bridges will not have to be built. When the Jhelum river bed level is brought below street level, floods will not occur. The dredged material could be used to strengthen the existing embankment besides raising a second line of defense bund at vulnerable spots.

09) Neglect of desilting a major contributing factor of present floods:

The J&K floods are as much a result of unprecedented rainfall over a week and also of neglect in desilting major rivers like Jhelum and Chenab, and their tributaries. The need to dredge the rivers increased because of growing soil erosion driven by deforestation, wild fires and encroachment in river catchments. A time bound program need to be chalked out to dredge out the Jhelum river from Khannabal to Khadanyar besides the flood spill channel. The Flood Master Plan must have considered all these aspects and after increasing the present carrying capacity of river and spill channel by desilting suggested another parallel spill channel to carry the balance flood discharge.

Between August 27 and September 3, 12 of the 22 districts received excessive rain, the IMD data shows. Ten of these, two in Jammu and eight in Kashmir, got more than double the normal rain. Srinagar was pounded by 373% more rain than normal for this time. Ganderbal got nine times more rain, Pulwama about 7 times and Anantnag five times more rainfall.

The torrential rain brought back memories of the Uttarakhand tragedy where 440% excess rainfall in June 2013 caused havoc. Over 5,700 died in the floods, most of them tourists. Like in Uttarakhand, other factors too are at work in J&K. One key reason for Jhelum's relentless rise is negligible desilting. An estimated 36 lakh cubic metres of silt has accumulated in the riverbed. Only recently, in 2012, two dredging machines were imported from the US to remove silt washed down by the river from the mountains. The last dredging was done in 1986. In 25 years, the meandering Jhelum got silted leaving little space in it to take excess water.

In 2009, the state's irrigation and flood-control department proposed a Rs 2,200 crore desilting project to the Centre. It included dredging of Jhelum's channels and anti-erosion work. Only Rs 97-crore portion of this for immediate action was approved. It included machines to dredge Jhelum, particularly of its floodspill channels in Srinagar and outflow streams at Baramulla. Work began late 2012. Experts say the river's carrying capacity is down from 45,000 cusecs (1975) to 32,000 cusecs (2012).One cusec is the flow of one cubic feet water in one second at any given point.

Illegal felling and forest fires contributed to soil erosion in catchments of Jhelum, Chenab and Tawi. Mountain streams washed down the loose soil that ended up in the river. Encroachments contribute to loss of soil cover. According to the recent statement made by the forest minister in the assembly in J&K, 14,345 ha of forestland is encroached upon - 9,463ha in Jammu and 4,878 ha in Kashmir.

10) Srinagar city Master Plan 2000-2021:

Problem of river discharge on river Jhelum had to be solved as it has assumed serious dimensions. Over the past fifty years, river Jhelum and spill channel has heavily silted up. It was understood that the flood control problem was being entrusted by I&FC Dept. to some consultancy firm; hence the problem of siltation, dredging, gradient, velocity etc. shall be dealt with. Some suggestions included to redesign flood absorption basin from Kandizal to Padshahibagh saving the railway line and Mahjoor nagar area. Beds of river Jhelum and spill channel were proposed to be deepened to increase the discharge capacity and ensure the minimum draft required for mechanized water transport. Gradient of spill channel was supposed to be increased up to the permissible limits. Weir at Chattabal had to be redesigned. Navigational channel of the lock gate at weir site was planned be desilted to make it functional. However the Srinagar city Master Plan 2000-2021 has also got entangled in the cobweb of red-tape like other similar vital issues, which ultimately land us into a chaotic situation like the recent floods.

11) *Srinagar Master Plan 2000-2021 recommendations for river Jhelum:*

- All Doongas and houseboats be shifted upstream of Padshahibagh or downstream of Chattabal. The tourist oriented houseboats could be shifted to Dal or Nigeen area.
- Encroachments made on the banks of the river and all three Khuls- Kuta Khul, Soner Khul and Watel Khul be cleared en-masse.
- Development of river fronts will involve clearance of some sites for development of parks.

-To stop the garbage-dumping from Lal-Ded Hospital; incineration plant be installed there.

-Wherever possible, Agriculture department should level the disused brick kiln sites on either side of Cement Bridge and use the same for growing vegetables.

Shikara ghats be constructed at appropriate points, connected with pucca stairs to nearby roads.

-While according permission for building construction on river and nallah fronts, no part of the building should protrude towards the river and nallah boundaries, or over their embankments.

(B) DISASTER MITIGATION / RESPONSE

In comparison to the instant response to the world disasters, the victims of the recent floods in Kashmir Valley feel that their State has been lacking in timely rescuing of the sufferers, besides sluggish arrangement of dewatering, their rehabilitation, provisions, shelter, clothing (*roti, kapda, makan*). The general public opinion is that the local youth swung into action instantly till army and other rescue teams mobilized their resources. For rehabilitation, despite tall claims, the State Govt. seems to have been let down by the Central Govt. in release of timely necessary funds required for the immediate relief measures of the flood victims. Every passing day is precious in view of the approaching harsh winter and it appears that people whose lives got spared from the wrath of floods may die now due to lack of proper warm shelters, warm clothing and other facilities. In the beginning there were tall claims of release of assistance in thousands of Crores of Rupees as relief to flood victims. The months of September and October have passed in waiting, but the promised USA's *seventh fleet of Bangladash did not come (*till the formulation of a new country became a reality). *Justice delayed is justice denied.* Today the business hub of Lal Chowk presents a devastated look with dust in the atmosphere and people buying the salvaged material from the footpath vendors, while the shops are getting renovated. Most of the houses that had submerged are getting cleared of the silt and the filth that had entered there in. Some of the damaged houses being unsafe are being dismantled in. There is an apprehension that new structures may come up with utter disregard to building bye-laws. SMC must exercise its check with the powers vested in it. Most of the recovered vehicles are lying near the workshops waiting for their turn as it takes number of days for restoration of each vehicle. In this connection most of the open spaces near workshops are fully occupied by the damaged vehicles waiting for their turn. Traffic jams seem to have increased despite stalling of a sizeable number of vehicles due to floods. The recovery of city to its original position may take a pretty long time.

From the natural disaster trends since 1945, scientists have observed that, there was an increase in the number of small and large natural disaster events during last half a century. Most of the natural disasters were meteorological in origin, river flooding, coastal inundation from hurricanes and typhoons, accounting them of a third of all disasters during this period of time. Earthquakes, although far fever in number are more fatal. As more than half deaths attributed to natural disasters resulted from earthquakes.

In J&K State Community Disaster Training was begun, wherein Divisional Disater Management Cell, in collaboration with Civil Defense Organization J&K State, Red Cross Society, Fire & Emergency Services and disaster administration participated in August 2013. Forty youth mostly from Dal area and other areas of Srinagar city attended the training program, whose services were proposed to be used as Civil Defense Volunteers. Later similar programs were reportedly held district wise to register maximum number of youth for any future eventuality. It was only after losing 1150 lives in six years of various natural disasters, besides 953 persons in Kashmir earthquake in 2005 and 192 persons in 2010 flash floods in Leh, that J&K State formulated Disaster Management Policy falling in line with other states like Andhra Pradesh, Assam, Arunachal Pradesh, Himachal Pradesh, Kerala and Manipur, who have put in place comprehensive plans to mitigate natural disasters.

However in the recent floods, local youth showed their valour and risked their lives to save people in distress. They made their own innovations by carrying the old sick people on inflated tyres of trucks placing cushions over carom boards. The authorities took time to mobilize their resources to save the remaining lot of people.

From the studies of the decade of environmental disasters of 1990's, widespread environmental degradation at global level floods, hurricanes, famines, windstorms, and other extreme events are becoming more frequent. Some important records of past historical tsunamis are:

- 1st November 1775- Earthquake caused tsunami destroyed Lisbon killing over 6000 people.
- 27th August 1883- Volcanic eruption tsunami from the Karakota volcano drowned 36,000 people.
- 15th June 1986- Tsunami in Japan killed over 27,000 people.
- December 26th 2004- Tsunami has beaten all records being the deadliest one till date, killing more than one lakh people and rendering several lakhs homeless. The strange observation made on the most-hit island of Indonesia was that animals had migrated from the affected areas and the aquatic life was not much damaged in Indian Ocean.

- After the Muzaffarabad earthquake in 2005, the US geologists RB Bilan and K.Wallace confirmed in a conference in India that the Kangra regions like other parts of the Himalayas are vulnerable to a future large earthquake of magnitude 8. Scientists warned that since the population in Gangetic basin in India and Pakistan is larger than at any time in the history, any future massive eartquake in Himalayas could have a much greater impact on population than the tsunami of 2004. Structural seismologists have warned that the Western Himalayas may be on a stress state similar to that of Andaman plate boundary prior to 2004. The view was however refuted according to their research and evidence by H. Gupta and Negi- former directors.

In May 1964 Yokahama Strategy emphasized that *disaster prevention, mitigation and preparedness is better than disaster response* in achieving the goals and objectives of vulnerability reduction. Disaster response alone is not sufficient as it yields only temporary results at a very high cost. Prevention and mitigation contribute to lasting improvement in safety and are essential in integrated disaster management.

The GOI has issued guidelines that where there is a shelf of projects, projects addressing mitigation wil be given a priority. It has also been mandated that each project in a hazard prone area will have disaster prevention / mitigation as a term of reference and the project document has to reflect as to how the project addresses that term of reference. According to these guidelines, Jhelum Master Plan Project must receive top priority in its approval from GOI.

It is reported that the measures for flood mitigation were taken from 1950 onwards. As against the total 40 million hectares prone to floods, areas of about 15 million hectares have been protected by construction of embankments. A number of dams and barrages have been constructed. The State Governments have been assisted to take up mitigation programs like construction of raised platforms etc. Floods continue to be a menace however mainly because of the huge quantity of silt being carried by the rivers emanating from Himalayas. The silt has raised the bed level of many rivers to above the level of country side. Embankments have also given rise to problems of drainage with heavy rainfall leading to water logging in areas outside the embankments as witnessed recently in Rajbagh, Jawahar nagar, Majoor nagar, Bemina etc. To evolve both short term and long term strategy for flood management / erosion control, Govt. of India has recently constituted a Central Task Force under the chairmanship of Chairman CWC. The task force will examine the causes of the problem of recurring floods and erosion in States and regions prone to flood and erosion and suggest short term and long term measures. The task Force was to submit its report in December 2004. Due to erratic behavior on monsoons, both low and medium rainfall regions, which constitute about 68 % of the total area, are vulnerable to periodical draughts. Our experience has been that almost every third year is a draught year.

Flood preparedness and response: In order to respond effectively to floods, Ministry of Home Affairs have initiated National Disaster Risk Management Program in all the flood prone States. Assistance is being provided to the States to draw up disaster management plans at the State, District, Block and Village levels. Awareness generation campaigns to sensitize all the stakeholders on the need of flood preparedness and mitigation measures. Elected representatives and officials are being trained in flood disaster management under the program. Bihar, Orissa, West Bengal, Assam and Uttar Pradesh are among the 17 multi-hazard prone States where this program is being implemented with UNDP, USAID and European Commission.

The other issues to be cared for are: Earthquake risk mitigation, National core group for ERM, Review of building bye-laws and their adoption, Development of revision of codes, Hazard safety cells in States, National program for capacity building of engineers and architects in earthquake risk mitigation, Training of rural masons, Earthquake engineering in undergraduate in engineering and architecture curricula, Hospital preparedness and emergency health management in medical education, Retrofitting of life line buildings, National earthquake risk mitigation project, mainstreaming mitigation in rural development schemes, National cyclone mitigation project, Landslide hazard mitigation, Disaster risk management program, Awareness generation, Disaster awareness in school curriculum, Information, education and communication, Special focus to Northern and Nort eastern states.

The various prevention and mitigation measures outlined above are aimed at building up the capabilities of the communities, voluntary organizations and Govt. functionaries at all levels. This is a major task being undertaken by the Govt. to put in place the mitigation measures for vulnerability reduction. This is just a beginning; the ultimate goal is to make prevention and mitigation a part of day-to-day life. The above measures would be put in place and the information disseminated over a period of five to eight years. The authorities are of firm conviction that with these measures in place, they could say with confidence that damages like those of recent past will not be allowed to recur in the country, at least at the cost, which the country has paid in terms of human lives, livestock, loss of property and means of livelihood.

07) The Flood Fury of 2014.

It is said that, *'Floods are acts of God, but acts of man cause flood damage'*. The recent floods of Kashmir Valley are a testimony to this fact. The Holy Quran states:
"We sent Noah to the people .... (With a message)
But they rejected him
And We delivered him and those with him
In the Ark
But We overwhelmed
In the Flood those
Who rejected Our Signs
They were indeed
A blind people!" (7:59-64)
Noah's warning was rejected by his generation and they were destroyed in the Flood. (C. 85)
The formation of *Satisar* is also reported to be a remnant of Noah's deluge. The Japanese scholars have recently expressed high regards for Kashmir as according to them it is the first land-mass to emerge after the floods of Prophet Noah (called Manu) receded. Lawrence quotes in his Valley of Kashmir that it is said that where the *Wullar* rests there was a great and a wicked city which was swallowed up by an earthquake, and the floods completed its destruction. The meaning of the word *'Wullar'* is cave and the legends say that the remains of the wicked city have been seen by the

boatmen. The formation of Dal Lake is also ascribed to the flooding of *Talni Marg* during the reign of Raja Parvarsen in sixth century AD, who constructed an embankment from Rainawari to Dalgate (now a road) to block the drainage of the newly formed lake. The Valley witnessed again major flood in 879 AD in the reign of king Awantiwarman, when the low-lying areas of the Valley were flooded due to blockade of river Jhelum down below Varmul and Er. Suya devised an ingenious method of removing the blockade by dropping gold coins in the river bed, which resulted into the clearance of debris by local divers followed by release of dammed up waters to push the blockade downstream. *"The flood of 1893 was a great calamity, but it had the good effect of warning the State that the valuable house property in Srinagar was inadequately protected. The protection works were taken in hand but at the same time, it was apprehended that the security of city means loss to cultivation on the banks of the river above Srinagar. The more Srinagar is protected the more obstruction there will be to passage of waters from south through the city. Thus, the founders of Srinagar have bequeathed a serious engineering problem to their successors", says Lawrence.* In 1959 floods, with almost equal discharge as of today, there has not been such a colossal damage as the pressure on the river was decreased first by allowing inundation of flood plains through Kandizal breach besides catering of one third of discharge by flood spill channel and also allowing a part of discharge to flow into Dal Lake, where water level was maintained lower than the present one, thus saving the city from inundation. In addition colonies had not come up at the low-lying areas of Rajbagh, Jawahirnagar, Mahjoor Nagar and Bemina etc., which formed flood lungs in emergencies. The Master Plan 2000-2021 describes that river Jhelum and its diversion channels namely Tsunti Khul, Kuta Khul, Soner Khul and Watel Khul were navigational per-se. These water courses contributed to a large extent to the environment, trade and water transport and helped to carrying down substantial volume of discharge during floods. Incidentally the proposed mechanized water transport on river Jhelum would have proved a great savior in the recent crisis. *"Water transport on the water courses has dwindled for the reason that over a period of time cross sections of river and khuls have squeezed, beds have risen and draft dropped down due to heavy siltation. The banks of the river and khuls have been mis-used by the public and encroached upon"*, says Lawrence.. Several recommendations have been made regarding reviving the carrying capacities of the river and the adjoining streams, but no action was taken till date despite passage of fourteen years of the city's Master Plan period. In earlier days it was a practice to mark with paint the HFL (Highest flood level) on the walls of important Govt. buildings to serve as a reference mark for raising the plinths of future constructions at least 0.75 meters higher than the HFL. It is high time that this practice is revived marking the current HFL for future guidance.

The recent flash floods have left many lessons for us to take care for future development. The chronological order of events that were reported in the media was as under:

01. Heavy rains lashed Jammu& Kashmir including summer capital Srinagar for the second consecutive day Wednesday triggering flood threat across the Valley. Water level at Ram Munshi Bagh in Srinagar 12 ft., six notches below danger mark; water level at Sangam Anantnag 21 ft., two notches below danger mark. Met. Deptt. forecast moderate to heavy rains to lash Jammu, Kashmir and Ladakh regions till Saturday morning. (Sept. 3rd. 2014)

02. Flood threat looms over Kashmir. (Sept. 4th.)

03. Kashmir floods throw life out of gear, several areas inundated, many structures damaged. CM reviews situation. Flood alert sounded. Water level in Jhelum touches record level of 31 ft. The discharge of Jhelum was 70,000 cusecs against normal discharge of 25,000 cusecs. A breach occurred at Kandizal area of Budgam. Authorities asked people living in flood-prone areas and embankments of rivers and

streams to shift to safer areas. More rains forecast on Friday. Flood waters breached many embankments in many low-lying areas in Kashmir including Srinagar, forcing people to move to safer places. Jhelum River crossed 30 ft. mark at Sangam in Anantnag- 7 ft. above danger mark. It touched 21.8 ft against the danger mark of 18 ft. at Ram Munshi Bagh. Rains inundate city center, residential colonies.

Met Deptt. said, *"though September is not a rainy season for Kashmir, but due to under- development of favourable weather system, there had been wide-spread heavy rain in past as well. One such year after 1980 was 1992 (September) when most parts of Kashmir received heavy rains apart from Sept. 1988 in Jammu region. In future also we cannot rule out heavy rainfall in September"*.

Srinagar received 88 mm. rain, Qazigund 286 mm.,Pahalgam 115 mm.,Kupwara 61 mm.,Kukarnag 219 mm., Jammu 107 mm., Banihal 248 mm.,Katra 158 mm., Badarwah 165 mm. and Gulmarg 139 mm. in past two days. Roads got damaged bridges washed away, villages got flooded in Anantnag, Kulgam, Pulwama, Ganderbal, Baramulla. National Highway closed. Educational institutions closed. Marriage invitations cancelled. CM reviews situation.

People helpless, Government sleepless. Doodganga bunds breached. Bone & Joint Hospital and residential colonies inundated. Bund breaches not plugged at Rawalpora, Peerbagh, Natipora, Chanapora. (Sept. 5th.) Telephones, Mobile phones, internet, Radio Kashmir, Door Darshan, electricity supply, water supply snapped. Press enclave submerged.

04. Flood fury, death toll 248. Twenty five bodies recovered from Srinagar. Doctors send alarm of epidemic. As floods recede, administration yet to come out of debris. Fear of dead bodies keeps people away from Jawahirnagar, locals complain of tardy dewatering. GK resumes publication after ten days.

05. Kashmir economy down by a trillion. Damage to infrastructure 100,000 cr. Houses either fully or partially damaged 300,000. Flood affected villages in Kashmir-1700, in Jammu 900. Roads damaged 12,553. Mobile and Internet not restored. Flooded Dal Lake tells its own tale of destruction, Lake dotted with ravaged houseboats, scary boat-wallas, reopening of civil secretariat proves damp squib. (Sept. 19th)

06. Toll mounts to 280. HC seeks Govt. response on 'tardy' relief measures. NGO's, Bill Gates announce relief for J&K floods. (Sept. 20th)

07. Day 14- Civil lines still submerged. Lal Chowk once a buzzing market turns into ghost- street. Scourging floods spawn tales of youth valor. (Sept. 21st)

08. 13 patients lost their lives as Govt. abandoned SMHS Hospital. Deluge destroyed Radiology Deptt., Medical ICU, ENT, Ophthamological facilities, Diagnostic labs etc.
Day 15-Thousands still out of their homes. Kashmir confronted devastating deluge with unity, compassion- uninterrupted relief, rescuers poured in from untouched areas. (Sept.22nd)

09. Day 16- Srinagar areas remain inundated. Kashmir inc cries criminal negligence demands probe. Deepening of river bed saved Ganderbal. Down town brave hearted rescued 300 people from flooded Lal Ded Hospital. (Sept.23rd)

10. Day 17- crises mounts in flood hit Srinagar. People fume as dewatering goes on at sluggish pace. Dewatering process goes awry, thanks to official apathy. CM meets PM demands special rehab package. (Sept. 24th)

11. Now JK seeks outside help to pump out flood water. SC panel to ascertain situation. 9 brave hearts, 2 boats and one rescue mission. Pampore youth brave flood fury to save 2000 people in 3 days. Daharmuna

swimmers saved 400 people in deluged Bemina. Volunteers executed 3-day operation with precision, rescued policemen, kids. (Sept.25th)

12. Prices of essentials sky rocket after floods. Govt. likely to submit loss memo to GOI by weekend. Water filters donated by Oxfam India struck in red-tape. Relief material unlikely to reach needy in view of hurdles created by J&K Govt. (GK Sept. 26th)

13. Devastating deluge- 12 lakh families hit in J&K. Kashmir boys extend helping hand from Bangalore to flood victims. (Sept. 28th)

14. Centre preparing comprehensive policy on Kashmir: Rajnath Singh. Pune's offer to help clean Srinagar found no takers. We offered support, were told to wait: Commissioner. Flood ravages JK's road infrastructure, Estimated damage Rs.1427 cr. (Sept. 29th)

15. Kashmir Floods- a disaster of international magnitude. Govt. clueless how Srinagar sank. Babus surface to defend cornered Govt. J&K inadequately prepared for floods. Flood havoc –PIL seeks probe into official negligence. (Sept. 30th)

16. JK delayed flood control project: CWC, SMHS Hospital out of service, Flood trauma can affect health of Kashmir children. (Oct 1st)

17. JK,CWC pass the buck over Jhelum flood project. In flood-hit villages, railway track, expressway 'sped up' devastation. (Oct.2nd)

18. Many more events got unnoticed or unreported in the media, a few instances are as:

i) Mr. Showkat a teacher in Fine Arts received an SMS at his home at Rainawari that flood waters in Jhelum have reached Rajbagh area. He rushed in a boat to save his wife, one month old son and his parents-in-law from Rajbagh locality. He rowed his boat over the bund along the current, boarded his family and others in the boat and rowed back now against the current, which was an uphill task for him. He saw three persons carried by the current near the bund and two persons drowned near the fountain outside Radio Kashmir building. Helplessly he could not save them. He saw a houseboat had been carried by the current upto TAO Café on the Residency road.

ii) Two officers of high profile along with their families were found rushing to airport in a motorboat, but got struck with an iron rod damaging the boat and were saved by a local of the area.

iii) A relative of ours under treatment was short of oxygen and was carried to SKIMS, thus his family escaped the wrath of floods, but he himself passed away, besides his house at Jawahirnagar crumbled down.

iv) Another relative on dialysis had to be lifted along with his family on a helicopter to carry him to airport and then to Delhi for safety.

v) Another promising boy who had invested everything in his business and owned a shop at Sangarmal shopping complex lost everything. Like that there must be innumerable happenings that got unreported.

vi) A family in Bemina lost their earning hand a few months back in a slip in his house, survived by a handicapped boy of 14 years, two small daughters, old aged mother-in-law and the ill-fated wife. They resided in a single storey house that got submerged and they shifted to the roof slab. Somehow they were rescued and walked over a kilometer up to Iqbal memorial crossing. They were provided shelter in a nearby two storey house, water followed them there too. Somehow after a great struggle they could be rescued after a five days.

Thus it is evident from above that both the public as well as Govt. were caught unawares in the flash floods both of whom never expected such an unprecedented wrath of flood waters. But the people charged with the task of flood protection, establishment of round the clock control room, organizing of yearly flood rehearsals, ensuring of alternative wireless communications, engineering the preparedness of the disaster management, ensuring instant relief measures, etc. cannot be absolved of their responsibilities. In fact Govt. is supposed to foresee and plan ahead for the upcoming events.

However it is but natural for water to overflow its banks in the event of rainfall in its upper catchment and spill into flood plains which are basically its right of way. Extensive and often unplanned use of flood plains, disregarding the basic fact, that it is a part and parcel of the river, leads to flood damage. Thus the uncontrolled and indiscriminate development of flood plains due to pressure of population can be considered as one of the main factors responsible for the ever increasing flood damage reported from the different parts of the country in spite of the substantial investment in the flood-sector during the last six decades.

Due to financial constraints no flood control structure can be constructed to provide total or absolute protection against all conceivable magnitude of floods. Moreover not all "flood prone" areas are amenable to protection through conventional flood-control measures due to a variety of reasons. For details and subjects of Flood Management, Concept of Flood Plain Zoning, Broad Methodology, Attempts in the past, Flood forecasting, Flood warning and the Valley Scenario, Engineering preparedness for disaster mitigation etc. please consult my book *Environment in Jammu & Kashmir* published2013 by M/S Gulshan Books Srinagar.

08) Kashmir Floods – a Chronology

While as volumes can be written on this topic; however what we need is time bound action plan before we face another such catastrophe as that of the last September floods, from which we may take a long time to recover.

اعوذ بالله السميع العليم من الشيطان ارجيم

بسم الله الرحمٰن الرحيم نحمد ه ونصلى علٰى رسوله الكريم

لقد ارسلنا نوحا الٰى قومه فقال يٰقوم اعبدالله مالكم من اله غيره انى اخاف عليكم عذاب يوم عظيم ة فكذبوه فانجيٰنه والذين معه فى الفلك واغرقنا الذين كذبوا بٰاياتنا انهم كانواقوما عمين ة

*We sent Noah to his people, He said, "O my people! Worship Allah! Ye have No other god but Him. I fear for you the Punishment of a dreadful Day! (7:59)/ But they rejected him, And We delivered him. And those with him. In the Ark: But We overwhelmed In the Flood those Who rejected Our Signs. They were indeed **A blind** people!(7:64)*

CHRONOLOGY OF FLOODS IN KASHMIR

About 40 million years ago Indian plate crashed into Eurasian plate at the geographically breakneck speed of 4 inches per year to form the Himalayas-

The collision created Himalayan Mountains welded together by warped and shattered rock interlocking to form the highest chain on earth. Out of this range of mountains, North Western Himalayas comprise of three states viz. J&K, Himachal Pradesh (HP) & Uttrakhund (UK), covering an area of about 33 million hectares, forming about 10 % of the total geographical area of the country. The region occupies the strategic position in the northern boundary of the nation and touches international boundaries of Nepal, China & Pakistan. Most of the area is covered by snow-clad peaks, glaciers of higher Himalayas & dense forest covers of mid Himalayas.

Kashmir Valley - Historical Background

Geologists believe that for *millions of years* Kashmir Valley remained under Tethya sea and the high sedimentary-rock hills seen in the valley now were once under water. Besides *a few million years* have passed when Kashmir Valley which was once a lake called Satisar came into its present form. It is also believed that Kashmir Valley was earlier affected by earthquakes. Once there was such a devastating earthquake that it broke open the mountain wall at Baramulla and the water of the Satisar lake flowed out leaving behind lacustrine mud on the margins of the mountains known as karewas. Thus came into existence the oval but irregular Valley of Kashmir. The karewas being in fact the remnants of this lake confirm this view. The karewas are found mostly to the west of the river Jhelum where these table-lands attain a height of about 380 meters above the level of the Valley. These karewas protrude towards the east and look like tongue-shaped spurs with deep ravines.

The history of Kashmir is dotted with occasional major floods and frequent minor floods every now and then. This is due to its mountainous topography which all forms its catchment area including that of its tributaries.

To begin with the formation of Satisar is also reported to be a remnant of Prophet Noah's deluge whereupon the Japanese scholars have recently expressed high regards for Kashmir as according to them it is the first land mass to emerge after the foods of Prophet Noah (AS) (called Manu) receded. Some archaeological finds have discovered structures that remained under water for a pretty long time, thus proving that habitation existed here before formation of Satisar. Traces of Himalayan ice age- (18,000 years ago the last ice age reached its height) and stone age – (Neolithic from 12,000 years ago) have been found in Kashmir as well.

- 4000 years flood history says that these floods were caused due to rains. Only two major floods were due to earthquakes.
- 2014 BC-In the era of Raja Sundar Sen (2083-2042 BC) – earthquake struck in the night time and old city of Sindmat Nagar sank underground, water gushed forth from bottom and Wullar lake came into being. Rock fell at Khadanyar Baramulla, the valley got drowned up to Bijbehara. Boatmen would see the roof tops under water for a long time. Budshah constructed Zaina Lank on the roof top of a structure.

- 7th century AD-Iin the era of Raja Durlab Dron (617-635 AD), river Jhelum breached its banks, it changed its direction at Nawpopra and entered the valley of Vital Marg and gave birth to Dal Lake.
- 8th century AD- In the era of Lalitadatya (715-752 AD), many week's rains covered whole city including Raj Mahal, which was shifted to Letapora; hundreds of houses were carried away by floods in Srinagar.
- 9th century AD-In Raja Avantiwarman's time (872-900 AD) an earthquake struck, rocks again came close at Khadanyar, entire area got drowned upto Bijbehara, causing famine. Soya devised an ingenious scheme of throwing gold coins in the river bed at Khadanyar. Divers cleared the way.
- 10th century AD- In the era of Raja Parth Warma(923-934 AD) floods carried houses of the city and dead bodies floated in the river.Paddy fields were destroyed causing famine.
- 12th century AD —In the era of Raja Harash Dev (1103-1114 AD) floods damaged all crops causing famines, people bought food stuffs by the weight of gold.
- 14th century AD- In the reign of Sultan Shahab-ud-Din (1360-1378 AD) floods damaged 20,000 houses in Srinagar, Sonawari and other low-lying areas.
- 16th century AD- In the era of Ali Shah Chak (1570-1579 AD) whole valley got inundated, all agricultural land was submerged, landslides took place, hundreds of houses got damaged, famine continued for three years.
- 17th century AD- In the era of Ibrahim Khan (1678-1686 AD)in the year 1683 there was continuous rain for one month causing devastating floods, houses were washed away which floated on water like boats with inmates weeping and wailing, all the bridges gave way, agriculture land and cattle were washed away in the floods. This was known as *"Tughyan-i-behad"* i.e. flood without borders. The areas that escaped floods were shaken by earthquakes, killing hundreds of people. Thousands of houses collapsed.
- 18th century AD- In the era of Nawazish Khan (1709-1710 AD) excess rains and winds caused floods resulting

in to great losses to agriculture and buildings. After this a devastating fire broke out in Mohalla Malchimar in Safakadal, which destroyed twenty ajacent Mohallas and 40,000 houses in them.

- Again in the era of Afrasiab Khan (1746-1748 AD) rains caused floods, damaging crops, river overflowed its banks, thousands of houses got damaged in the city, people died of starvation, the dead bodies could not be handled, shrouds were rare. Dead bodies would be wrapped in grass and thrown into river, which contaminated the water. About one third population perished, others fled the country and the rest stayed back to face the famine.

- In the era of Amir Khan (1771-1772 AD) floods hit the country and his Diwan Khana was washed away along with much of agricultural land and all the bridges. After the floods Sher Garhi was rebuilt with strong walls and grand buildings.

- 19th century- in the era of Shaikh Ghulam Mohi-ud-Din in 1841 AD, Jhelum overflowed its banks due to incessant rains. There was a breach of Qazizad bund and water entered Srinagar. Maximum damage took place in Rainawari and Khanyar areas. All bridges from Fatehkadal to Sumbal were washed away.

- 20th century AD- In the era of Maharaja Partap Singh there was rain for 59 hours on 24th July 1903, Jhelum banks overflowed, all the low lying areas of the city were flooded. People saved their lives by rushing to higher areas. Houses and cattle got washed away. Six out of the seven bridges on river Jhelum got washed away by piston like action of the gushing flood waters. Many people died because of drowning. British Engineers suggested channelizing the river Jhelum by raising embankments on either side. However Lawrence has been skeptical about it saying that by raising the banks they are leaving a severe problem for the posterity to contain the natural right of way of the river Jhelum, which came true after a century of his prediction.

- In 1959 AD in the era of Bakhshi Ghulam Mohammad, flood known as FORD damaged crops and property.

- The recent floods during 5th September to 10th September 2014 when Umar Abdullah was the CM, whole of J&K State got covered with it gradually. It was the worst flood according to its intensity and spread area. Almost all the districts of the State were hit by it. In this flood rich and poor witnessed death closely shaving and following them and their property perishing before their very eyes. We must take lessons from this flood which made people to realize the unreliability of the worldly wealth. The need of the hour is to tie our belts to move forward with patience and steadfastness and pray to Almighty Allah to widen the gates of mercy and benefits for us.

- Our ancestors have been braving these floods right from the day one. Though there have been efforts made earlier to contain the floods for saving the capital city and the measures adopted have stood their test of time for the past about a century, but it is unfortunate that the attention of authorities have never gone to seriously devise means to safeguard the lives and property for the times ahead. In fact Governments must foresee and plan in advance, for which past history provides them lessons to learn.

- *If Netherlands can remain safe below the sea level, why can't we protect our valley from the catastrophe that has been striking us every now and then.*

- The recent floods have provided us an opportunity to rise to the occasion and give top priority to first rehabilitation of the flood victims and next to reform the Master Plan for floods on the advice of the expert consultants equipped with latest technologies of remote sensing etc. and adopt controlling measures without loss of any time. There is no guarantee that a similar catastrophe may not hit us soon again particularly due to global climate change.

Within seven months this has been followed by another flood in March 2015 giving sleepless nights to many besides causing landslides. It is now over eleven months and with a few hours of rain people get alarmed and the erratic behavior of weather aggravates the situation with frequent cloud bursts.

According to Lawrence Kashmir valley has witnessed many calamities like floods, famines, fires, epidemics, earthquakes, forced labor (begaar) in the 20th century that has taken toll of many lives and caused distress to people and made them skeptical about imminent calamities. As he states in "Valley of Kashmir" that a person who is beaten up on a roadside for no fault of his loses his own respect and that of others." That was the state of affairs of people then and not much has changed since then as we are now victims of bullets for the past quarter of a century. He also states that "The condition would have improved had there been a strong rule for two generations", which has never come nor is expected ever to come in near future. However there has been a silver lining, a significant and drastic change in the thinking and action of the present day youth as compared to those times, when no person moved to fetch a pail of water from the river Jhelum to extinguish fire that devastated half of the city of Srinagar, the people assumed that it was a wrath of God, hence *"the wretched Kashmiri did not act to combat the calamity"* as stated by Lawrence. Thanks to our local youth who have come to the instant rescue of the flood victims of September 2014, at the risk of their own lives. We salute their courage and bravery. They have proved worthy of Dr. Iqbal's conviction about Kashmiris :

جس خاک کے ضمیر میں ہو آتش چنار ممکن نہیں کہ سرد ہو وہ خاک ارجمند

At the same time the worst ever flood of September followed by another jolt in March 2015 just after seven months has shaken the conscience of the erring public who have been abusing the river by encroachments on its flood absorption basins and the authorities who have allowed it to happen before their very eyes. The civil society feels strongly to rise to the occasion to bring a check of this ongoing vandalism that may one day drown us all due to our faults.

Reconstruction of the valley:

Reconstruction of the flood-prone devastated city areas:

Now that the centuries' worst flood has receded, people shall be /are rebuilding their damaged houses in their respective sites.

Some important points need to be considered to prevent recurrence of such an eventuality in future.

- The immediate challenge was to provide them food, warm clothing and shelter *(Roti, Kapda, Makan)* to escape from the approaching winter, besides till the time their structures are re-erected. Prefab structures or rented accommodation was proposed.

- There is no guarantee that floods of even worse magnitude may not hit the valley again. The global climate change is one of the major factors.(*This apprehension has proved right in March 2015 just after 7 months only followed by frequent cloud bursts intermittently.*)
- Government seems to be contemplating to provide a parallel spill channel on the upstream side of the existing one, but that may be a long drawn affair as it shall involve time consuming land compensation, construction of many bridges, aqua ducts etc. Besides this channel may also get defunct with the passage of time like the existing one for want of its regular maintenance.
- De-silting of the river bed from Khannabal to Khadanyar and flood spill channel with sufficient number of dredgers to increase the carrying capacity of the river and the spill channel particularly in winter months needs to be taken up forthwith, before onset of next flood season. The excavated material could be used for making a second line of defense bund behind the existing one.
- Construction of a second line of defense bund parallel to the present one at weak spots could be considered to save the low lying areas from inundation.
- The other ideal solution would have been to plan a new city on higher contours along foot-hills from Ganderbal to Harwan, Pandrethan, Khunmoh to Rajpora and on karewas along southern foot-hills ensuring safety from river floods, proper drainage, guaranteed sunlight and fresh mountain breeze. The recently flooded areas could be allowed to form a mini lake but again that may be a herculean task.

- People in present distress cannot wait for the long term measures of flood protection and they will soon/have resortted to the reconstruction of their collapsed/damaged houses.

- SMC must come up with the new building norms that are required in the flood prone areas to ensure the safety of lives of the inhabitants.

- New constructions could come up as framed structures. Basement floors could be raised on RCC columns with ceiling level higher than the HFL, leaving the space for car parking etc. In fact such norms that already exist for shopping complexes in Srinagar city have been violated causing parking problems in the city. SMC must implement these norms strictly.

- Safe foundations need to be designed as per BIS specifications after conducting soil tests, monitored by SMC. In fact I found a ten story structure was stalled by Muncipal Authorities at plinth level in Abu Dhabi, the reason being provision of lesser steel than the approved/designed one.

- Due to land hunger possibility of vertical expansion as already advocated, for the city be considered seriously to tackle the future housing needs in view of multiplying population besides relocating the families of flood hit/prone areas.

- The partially damaged structures of the flooded area need to be inspected by an expert team to suggest measures for their retrofitting.

- Dewatering stations need to be lifted higher than the HFL of 100 year flood for making these functional during crisis.

- The plinths of all Govt. buildings/establishments in the flood prone areas need to be raised higher than the HFL. The basement floors could be used for car parking etc.

- Flood plain zoning is useful in reducing the damage caused by drainage congestion particularly in urban area where on grounds of economy and other considerations urban drainage may not be designed for the worst possible conditions and presupposes some damage during storms whose magnitude exceeds that for which the drainage system is designed.

- The steps involved in implementation of flood plain zoning measures could be as:

1. Demarcation of areas liable to floods.
2. Preparation of detailed contour plans of such areas to a large scale (preferably 1:5000) showing contours at interval of 0.3 to 0.5 meters.
3. Fixation of reference river gauges and determination of areas likely to be inundated for different water levels and magnitudes of floods.
4. Demarcation of areas liable to flooding by floods of different frequencies like once in two years, five, ten, twenty, fifty and hundred years. Similarly areas likely to be affected on account of accumulated rainfall like 5, 10, 25, and 50 years.
5. Delineation of the types of which the flood plains can be put to in the light of © and (d) above with indication of safeguards to be ensured.

In the existing developed areas possibilities of protecting /relocation/exchanging the sites of vital installations like electricity substations/powerhouses, telephone exchanges etc. should be seriously examined so that these are always safe from possible flood damage. Similarly the pump stations of tube wells for drinking water supply should be raised above the HFL corresponding a 100 year flood.

Similarly possibility of removing buildings/structures obstructing existing natural drainage should be seriously considered. In any case unplanned growth shall be restricted so that no constructions obstructing natural drainage resulting in increased flood is allowed. In future the following regulations may be stipulated:

1. Plinth levels of all buildings should be nearly 0.75 to one meter above the drainage/submersion levels.
2. Damp proof course with asphalt course be provided at plinth level for masonry structures.
3. In the areas liable to floods all the buildings a stairway should invariably be provided to the roofs/attic floors so that temporary shelter can be taken there. The roof levels

of the single story buildings and the first floor level in double story buildings should be above flood level of 1 to 100 frequency so that the human lives and the movable property can take temporary shelter there when necessary during the floods.

In the past CWC prepared guidelines in 1873-74 for flood plain zoning which were approved by Central Flood Control Board. CWC also prepared a model draft and circulated it in the Ministry of Irrigation in 1975, to all the states for enacting legislature. However the response from states except Manipur has not been encouraging. Manipur enacted a legislation in 1978 which came into force in 1985.

Flood proofing measures, help greatly in the mitigation of distress and provide immediate relief to the population in flood prone areas. It is essentially a combination of structural change and emergency action, not involving any evacuation. The technique adopted consists of providing raised platforms for flood shelter for men and cattle and raising the public utility installation above flood levels.

In case of urban areas, certain measures that can be put into action as soon as a flood warning in received involve:

Installation of removable covers such as steel or aluminum bulk heads over doors and windows or other openings keeping stone counters on wheels, closing of sewer well, anchoring machinery, covering machinery with plastic sheet, seepage control etc.

Life jackets, inflated boats need to be stocked by the flood prone households. The youth need to be taught swimming to face any such future eventuality.

Flood proofing also tends to encourage persistent human occupation of flood plains.

Out of the non-structural measures "flood forecasting and warning" is considered as one of the most important, reliable and cost effective methods. CWC organizes flood forecasting at 157 stations in the country, of which 132 are for water stage forecast and 25 for inflow forecast for certain major reservoirs. The Flood Meteorological Offices (FMO) also provide information regarding general meteorological situation, rainfall of last 24 hours for different regions and range of quantitative precipitation forecasts for various river basins to the respective flood forecasting centers of CWC. All the data is simultaneously transmitted to the circle headquarters supervising forecasting works for overall security, monitoring, analysis and compilation. The final forecasts are then transmitted to the administrative and engineering authorities of the state and other user agencies connected with flood protection and management work on telephone or by special messenger/ telegraph/ workers depending upon local factors like vulnerability of the area and availability of the communication facility etc.

TWO DAY SEMINAR ON FLOODS OF 2014 HELD AT LALIT   GRAND PALACE HOTEL IN NOVEMBER-14

A two day national Seminar on "Retrospective and Prospective of 2014 Kashmir Floods for Building Flood Resilient Kashmir" was held at Srinagar from 15-16 November 2014. The Seminar was organized jointly by the Department of Earth Sciences, Kashmir University and Centre for Dialogue and Reconciliation (CDR). The 2014 flood was triggered by the complex interplay of atmospheric disturbances that brought widespread and extreme rains all across the state. The Jhelum waters, that used to be the provider of life and sustenance, suddenly became a monstrously destructive force against the human life and the infrastructure that cohabit its backyards since millennia.

Any future flood strategies for Jhelum Basin shall benefit from our learning from this horrendous experience and the threadbare deliberations held at the two days National Seminar. The September 2014 floods were unprecedented in the flood history of Kashmir and got everyone concerned about the consequences of another such disaster if it recurred. The immediate steps to be taken are to develop a strategy for mitigating floods in the state and that requires realization across the region at the local and the national level. The aim of this symposium was to conduct deliberations with the select group of relevant people who have the expertise to recommend and formulate a long term action plan for flood disaster management and mitigation in the state of Jammu and Kashmir.

The experts from various central agencies (Central Water Commission-CWC, National Institute of Hydrology-NIH,

National Geophysical Research Institute-NGRI, Central Groundwater Board-CGB, National Disaster Management Authority, NRSC/ISRO and National Green Tribunal), India Meteorological Department (IMD) and State Government agencies – Irrigation and Flood Control (IFC), Public Health Engineering (PHE), Rural Development, LAWDA, Srinagar Development Authority-SDA, IMPA, Agriculture Department, academia from Kashmir University, Indian Institute of Technology-Roorkee-IIT, National Institute of Technology - NIT-Srinagar, Jammu University, and various segments of the civil society, including experienced professionals, attended the Seminar.

Short-term and Urgent Recommendations:

The following recommendations made at the Seminar need to be taken up on priority immediately and could be accomplished in the shortest possible time to reduce the risk to the public and to property in Jhelum Basin from flooding.

01) Knowledge driven all-inclusive multidisciplinary flood planning needs to be initiated on priority by engaging technocrats with relevant expertise to develop insights into flooding mechanisms in the Jhelum Basin building on comprehensive existing studies.

02) Strengthening the flood infrastructure in the Jhelum Basin to cope up with the probability of next extreme flooding event of the magnitude observed in 2014. This includes the preparation of an integrated DPR for the construction of the alternate flood channel from Dogripora to Wullar, increasing the carrying capacity of the main Jhelum, dredging of the existing flood channel, dredging of the wetlands like Hokersar, Narkara, Nowgam Jheel, and Wullar lake, and strengthening of breached and weak embankments the broad plan of which is before the CWC.

03) The management of the water bodies/lakes and wetlands in the Jhelum Basin needs to be brought under one regulatory authority for their integrated management, being a single catchment area served by the same watershed.

04) The government, with the help of academia/research institutes, must consider undertaking a scoping study to assess

the probability of flooding in immediate future based on the understanding to be developed from the interactions of Ground Water, surface water and the glacier-melt in the Jhelum Basin.

05) Urgently operationalising the Flood Early Warning System (FEWS) for Jhelum and Chenab.

06) The State Government must initiate on priority (with the help of leading academic institutions), to undertake transparent flood-zonation and flood vulnerability assessments of people and places at village level so that the flood risk reduction is integrated with developmental planning at village level in all District Development Plans.

07) Government consider assigning proposals for bringing the technical ingenuity of the Irrigation & Flood Control in operationalising of FEWS, basin wide IFM and flood scenario mapping. The identified scientific studies on various aspects of flooding identified above are required to be undertaken on priority by involving Universities, consultants and institutes both national and international.

Urgent Long-term Recommendations:

The following recommendations made at the Seminar need to be initiated immediately and might take a few years to complete for flood risk reduction in the Jhelum Basin.

01) There is an urgent need to institutionalize the disaster management in the state by setting up of a vibrant and structured State Disaster Management Authority with a clear mandate to build the capacity of the state to prepare for, protect against, respond to, recover from, and mitigate all types of hazards, the state is vulnerable to.

02) Strengthening of flood control infrastructure in 4 high gradient streams in the south Kashmir viz., Rambiara, Veshu, Romshi and Lidder, that enormously contribute to the discharge at Sangam using the available techniques, so that the flood peak and concentration time is appreciably delayed by staggering them in the watershed itself before their discharge into the Jhelum at Sangam.

03) Initiating a massive capacity building program for building public awareness and soliciting public involvement in flood risk reduction.

04) In order to arrest the siltation of the watercourses from the catchment, the participants recommended the massive reforestation of the Jhelum catchment under CAMPA, IWMP and other existing governmental schemes.

05) Structural and non-structural erosion control measures in the high gradient tributaries in the south Kashmir viz., Rambiara, Veshu, Romshi, Lidder, Bringi and Aripath.

06) Consolidation of the fragmented data and knowledge into a database so that it is available to everybody for use on understanding the hydrological and meteorological processes and phenomena in the state.

07) Strictly regulating mining of the riverbed keeping in view the river/channel morphology and other required hydrologic and geologic criteria.

08) The flood disaster preparedness at government and community levels need to be strengthened so that there is a well-rehearsed mechanism in place for quick response, despite all the adversities and limitations, to minimize the impacts of flooding on the people and property.

09) Revision of the existing land use policy and building codes is required and enforce strict implementation in order to minimize human and economic loss in the event of natural disaster.

10) Comprehensive community based Disaster risk reduction plans need to be prepared on priority.

Long-term Recommended Measures:

These long term recommended measures are essential for building the necessary flood control

Infrastructure in the basin so that in the eventuality of the next extreme flood event, the loss of life and property is reduced appreciably in the basin

a) Construction of the alternate flood channel from Dogripora to Wullar

b) Improving the drainage system in the urban areas of the Jhelum Basin including the restoration of natural drainages wherever possible

c) The government needs to initiate programs aimed at conservation and restoration of the degraded wetlands in the Jhelum Basin to enhance their flood mitigation, in selected cases even sewage treatment functionality.. Bring city and town planning in the state into consonance with the flood and earthquake vulnerability.

d) Structural and non-structural measures be initiated under the supervision of I&FC for erosion control in the central and north Kashmir part of the Jhelum basin.

The Master Plan 2000-2021 recommendations for river Jhelum were:

- All Doongas and houseboats be shifted upstream of Padshahibagh or downstream of Chattabal. The tourist oriented houseboats could be shifted to Dal or Nigeen area.
- Encroachments made on the banks of the river and all three Khuls- Kuta Khul, Soner Khul and Watel Khul be cleared en-masse.
- Development of river fronts will involve clearance of some sites for development of parks.
- To stop garbage dumping from Lal-Ded Hospital, incineration plant be installed there.
- Wherever possible Agriculture department should level the disused brick kiln sites on either side of cement bridge and use the same for growing vegetables.
- Shikara ghats be constructed at appropriate points, connected with pucca stairs to nearby roads.
- While according permission for building construction on river and nallah fronts, no part of the building is protruded towards the river and nallah boundaries or over their embankments.
- Problem of river discharge on river Jhelum be solved as it has assumed serious dimensions. Over the past fifty

years, river Jhelum and spill channel has heavily silted up. It was understood that the flood control problem was being entrusted by I&FC Deptt. to some consultancy firm, hence the problem of siltation, dredging, gradient, velocity etc. shall be dealt with. Some suggestions included to redesign flood absorption basin from Kandizal to Padshahibagh saving the railway line and Mahjoor nagar area. Beds of river Jhelum and spill channel be deepened to increase the discharge capacity and ensure the minimum draft required for mechanized water transport.

- Gradient of spill channel be increased up to the permissible limits. (*which seems to have been shelved by the Govt. after a great hype*)
- Weir at Chattabal be redesigned. Navigational channel of the lock gate at weir site be desilted to make it functional.

However the Master Plan 2000-2021 has also got entangled in the cobweb of red-tape like other similar vital issues which ultimately land us into a chaotic situation like the recent one.

Disaster Mitigation:

From the natural disaster trends since 1945, scientists have observed that, there was an increase in the number of small and large natural disaster events during last half a century. Most of the natural disasters were meteorological in origin, river flooding, coastal inundation from hurricanes and typhoons, accounting them of a third of all disasters during this period of time. Earthquakes, although far fever in number are more fatal. As more than half deaths attributed to natural disasters resulted from earthquakes.

In J&K State Community Disaster Training was begun, wherein Divisional Disater Management Cell, in collaboration with Civil Defense Organization J&K State, Red Cross Society, Fire & Emergency Services and disaster administration participated in August 2013. Forty youth mostly from Dal area and other areas of Srinagar city attended the training program, whose services were proposed to be used as Civil Defense Volunteers. Later similar programs were reportedly held district wise to register maximum number of youth for any future eventuality. It was only after losing 1150 lives in six years of various natural disasters, besides 953 persons in Kashmir earthquake in 2005 and 192 persons in 2010 flash floods in Leh, that J&K State formulated Disaster Management Policy falling in line with other states like Andhra Pradesh, Assam, Arunachal Pradesh, Himachal Pradesh, Kerala and Manipur, who have put in place comprehensive plans to mitigate natural disasters.

The other issues to be cared for are: Earthquake risk mitigation, National core group for ERM, Review of building bye-laws and their adoption, Development of revision of codes, Hazard safety cells in States, National program for capacity building of engineers and architects in earthquake risk mitigation, Training of rural masons, Earthquake engineering in undergraduate in engineering and architecture curricula, Hospital preparedness and emergency health management in medical education, Retrofitting of life line buildings, National earthquake risk mitigation project, mainstreaming mitigation in rural development schemes, National cyclone mitigation project, Landslide hazard mitigation, Disaster risk management program, Awareness generation, Disaster awareness in school curriculum, Information, education and communication, Special focus to Northern and North eastern states.

The various prevention and mitigation measures outlined above are aimed at building up the capabilities of the communities, voluntary organizations and Govt. functionaries at all levels. This is a major task being undertaken by the Govt. to put in place the mitigation measures for vulnerability reduction. This is just a beginning; the ultimate goal is to make prevention and mitigation a part of day-to-day life. The above measures would be put in place and the information disseminated over a period of five to eight years. The authorities are of firm conviction that with these measures in place, they could say with confidence that catastrophes like those of recent past will not be allowed to recur in the country, at least at the cost, which the country has paid in terms of human lives, livestock, loss of property and means of livelihood.

However in the recent floods in absence of all these measures, it was the local youth who showed their valor and risked their lives to save people in distress. They made their own innovations by carrying the old sick people on inflated tyres placing cushions over carom boards over these besides using empty water tanks.

The authorities took time to mobilize their resources to save the remaining lot of people.

09) Short Term / Long Term Measures Needed to prevent Devastation by Floods

The media report after the September floods stated that "The Jammu and Kashmir government said the State suffered losses of Rs 1 trillion in the floods and 12.5 lakh families were affected.

According to preliminary estimates, the housing sector suffered losses over Rs 30,000 crore while business sector incurred losses worth over Rs 70,000 crores," J&K Chief Secretary Mohammad Iqbal Khanday said while addressing a press conference here.

Maintaining that State had never witnessed such a disaster before, he said this (flood) was not a disaster of national but international ramifications. "The State has never recorded such flood level in the past".

He said as per the initial assessment reports on the damages to the private property, a total of 3,53,864 structures have been damaged. "83,044 pucca houses have been fully damaged and 96,089 partially. Similarly, 21,162 kachha houses were fully damaged and 54,264 partially damaged besides 99,305 huts, cowsheds, were also damaged".

Mr.Khanday said 12.5 Lakh families were affected by flood across the State.

"281 lives have been lost due to floods in the State. 196 people lost their lives in Jammu province and 85 in Kashmir," he said adding 29 persons are still missing.

Many areas in Srinagar including posh localities of Rajbagh, Jawahar Nagar and Indira Nagar are still under water and flood water has not been drained out completely in last 22 days.

Mr.Khanday said 5,642 villages were affected by the flood across the State with 2,489 in Kashmir and 3,153 in Jammu division. "At least 800 villages remained submerged for over two weeks".

Referring to the damage caused to roads and bridges, Chief Secretary said, "Over 550 bridges/culverts were damaged. Besides, 6,000 km road network was also damaged by the flood water".

On disposal of carcasses of livestock, he said, "Over 1500 carcasses have been removed from Srinagar only and scientifically disposed off. Besides, hundreds of tons of garbage is being removed from the city on daily basis".

Asked whether there was any outbreak of any disease in the flood-hit areas," Mr. Khanday said, "No outbreak of any disease has been reported from anywhere in the State."

"The situation is being constantly monitored and 7 lakh children in the age group of 6 months to 15 years have been vaccinated"

A) *Desilting of Jhelum yet to take off (Kashmir Monitor-29 the April 2015)*

Even if a slight rainfall causes the water level to cross the danger mark in Jhelum and revives the dreaded memory of September deluge in public psyche, the Irrigation and Flood Control department is yet to start the "much needed" dredging of the river to increase its carrying capacity, thanks to the lack of approval from the union government.

Official sources told The Kashmir Monitor the IFC department has already sent the proposal to de-silt Jhelum and was expecting to receive the final approval for work andFUNDS earlier this month but the Ministry of Water Affairs has failed to take any decision so far. "We are awaiting the central ministry approval to start the machine dredging of river. The technical advisory committee of Ministry of Water Affairs could not take a decision on the 7th of the month. They will fix a new date and as soon as we receive the approval we will start the work. They have almost approved it but the final approval andFUNDS ⟳ are awaited," Chief Engineer, IFC Javed Jaffar said adding it was a top priority.

"Manual dredging of the river has already begun in some places like Anantnag, Bijbehara but that is possible only when the water level in Jhelum is less. We are hopeful the government would soon approve the proposal of full-fledged dredging of the river so we can start the de-siltation using machines," Jaffar said.

A report prepared by the department of Environment and Remote Sensing in the aftermath of September 2014 floods had recommended immediate de- siltation programme both for river Jhelum and its tributaries. Officials say after de-siltation, the main Jhelum river can cater to **45,000** cusecs from current **25000-30000** cusecs.

The government is already mulling to construct a new flood channel from Dogripora to Wullar to carry the surplus flood discharge of Jhelum that is expected to cost over **Rs 18,000 crore**. The official report in the aftermath of September floods had also recommended de-siltation of wetlands across Kashmir.

Experts believe de-siltation in the form of sand mining was the main reason that north Kashmir districts remained unaffected during the September foods. The carrying capacity of outfall channel of River Jhelum from Wullar to Khadinyar Baramulla has been increased by way of excavating **4,00,000** cubic metres of sand since April 2012, according to official documents.

Kashmir again faced prospect of floods last month as incessant rainfall increased the water level in Jhelum causing it to cross the danger mark. The government is yet to remove the silt from Jhelum that had deposited in Jhelum during September floods.

It is now July 2015, eleven months have elapsed and the proposed action to face the imminent floods is yet to begin. We have witnessed that most of the Govt. works are taken in hand in anticipation of allotment of funds in case of emergency, besides also issuing short term tenders or even work orders. An emergency work that endangers lives and property of a large section of people is delayed for unknown reasons. With this state of affairs, people are losing faith in the present set up as justice delayed is justice denied. Same is the case with the promised timely distribution of relief to the flood victims who continue to suffer. It was reported that some scanty relief cheques were distributed among some persons which reminds us about the couplet:

Samandar se mile pyase ko shabnam; khudaya ye bakhili hay ki razaqi.

Besides the much hyped promised aid is yet to come even after 11 months, that reminds us of the promised American seventh fleet to help Pakistan in Bangladesh crises, that never came.

B) Reconstruction of the Valley

Reconstruction of the flood-prone devastated city areas – these suggestions were presented by me in a seminar on

November 9, 2014 organized by Sakhawat Center at Lala Rukh Hotel Srinagar.

Now that the centuries' worst flood has receded, people shall be rebuilding their damaged houses in their respective sites. Some important points need to be considered to prevent recurrence of such an eventuality in future.

- There is no guarantee that floods of even worse magnitude may not visit the valley any time again. The global climate change is one of the major factors of this erratic behavior of weather. With the saturated subsoil combined with melting snows, a few hours of frequent torrential rains raises alarm and the affected people become restless with the apprehension of repetition of the last September devastating floods. With the flood absorption basins like Bemina, converted to housing colonies after resorting to an earth fill of over 7 feet and with the choked river cross-section by its beautification measures combined with construction of new bridges in close vicinity, water is bound to overflow its banks. Unfortunately the technical opinion is generally overrlued by a non-technical bureaucrat upheld by a Minister resulting into the catastrophe.
- Government seems to be contemplating to provide a parallel spill channel on the upstream side of the existing one, but that may be a long drawn affair as it shall involve huge and time consuming land compensation etc. Besides this channel may also get defunct with the passage of time like the present FS channel, for want of its maintenance.
- The other ideal solution would have been to plan a new city on higher contours along foot-hills from Ganderbal to Harwan, Pandrethan, Khunmoh to Rajpora and on Karewas along southern foot-hills ensuring safety from river floods, proper drainage

and fresh mountain breeze but again that may be a herculean task.

- Immediately the weak spots of the embankments need to be strengthened particularly at curve points, where flood waters hit most. Raising of RCC protection walls need to be considered at such spots.

- Instead of raising the levels of embankments, time bound dredging and de-silting of the river bed, FS Channel and flood absorption basins need to be resorted to without any loss of time and the process should continue throughout the year as siltation is a regular feature during the floods and in normal flows as well. For this purpose sufficient number of suitable dredgers need to be procured and deployed at pre determined spots to ensure dredging of river from Khannabal to Khadanyar and that of FS Channel besides other wetlands/other water bodies. This process could have been started eleven months back, but the authorities seem to be awaiting another disaster before they will wake up to the situation.

- Forestation of the catchment area with construction of check dams needs to be taken up in a planned manner.

- People in present distress cannot wait for the long term measures of flood protection and they will soon resort to the reconstruction of their collapsed/damaged houses.

- Govt./SMC must come up with the new building norms that are required in the flood prone areas to ensure the safety of lives of the inhabitants.

- Basement floors could be raised on RCC columns with ceiling level higher than the HFL, leaving the space for car parking etc. In fact such norms already provided for shopping complexes have been violated causing parking problems in the city center. SMC must implement these norms strictly.

- Life jackets and inflated rubber boats should be stocked by all the house-holds falling in the flood prone area. Young boys/girls need to be taught swimming to face any eventuality.
- Safe foundations need to be designed as per BIS specifications with the approval /check of SMC. In fact I found a ten story structure was stalled by Muncipal Authorities at plinth level in Abu Dhabi, the reason being provision of lesser steel than the approved/designed one.
- Due to land hunger possibility of vertical expansion of the city as already advocated by me in GK write ups, be considered seriously to tackle the future housing needs in view of multiplying population besides relocating the families of flood hit/prone areas.
- The partially damaged structures of the flooded area need to be inspected by an expert team to suggest measures for their retrofitting.
- Dewatering stations need to be lifted higher than the HFL of 100 year flood for making these functional during crisis.
- The plinths of all Govt. buildings/establishments in the flood prone areas need to be raised higher than the HFL. The basement floors could be used for car parking etc.
- Flood plain zoning is useful in reducing the damage caused by drainage congestion particularly in urban area where on grounds of economy and other considerations urban drainage may not be designed for the worst possible conditions and presupposes some damage during storms whose magnitude exceeds that for which the drainage system is designed.
- The steps involved in implementation of flood plain zoning measures could be as:
1. Demarcation of areas liable to floods.

2. Preparation of detailed contour plans of such areas to a large scale (preferably 1:5000) showing contours at interval of 0'3 to 0.5 meters.

3. Fixation of reference river gauges and determination of areas likely to be inundated for different water levels and magnitudes of floods.

4. Demarcation of areas liable to flooding by floods of different frequencies like once in two years, five, ten, twenty, fifty and hundred years. Similarly areas likely to be affected on account of accumulated rainfall like 5, 10, 25, and 50 years.

5. Delineation of the types of which the flood plains can be put to in the light of © and (d) above with indication of safeguards to be ensured.

In the existing developed areas possibilities of protecting /relocation/exchanging the sites of vital installations like electricity substations/powerhouses, telephone exchanges etc. should be seriously examined so that these are always safe from possible flood damage. Similarly the pump stations of tube wells for drinking water supply should be raised above the HFL corresponding a 100 year flood.

Similarly possibility of removing buildings/structures obstructing existing natural drainage should be seriously considered. In any case unplanned growth shall be restricted so that no constructions obstructing natural drainage resulting in increased flood is allowed. In future the following regulations may be stipulated:

1. Plinth levels of all buildings should be nearly 0.75 to one meter above the drainage/submersion levels.

2. In the areas liable to floods all the buildings a stairway should invariably be provided to the roofs/attic floors so that temporary shelter can be taken there. The roof levels of the single story buildings and the first floor level in double story buildings should be above flood level of 1 to 100 frequency so that the human lives and the movable

property can take temporary shelter there when necessary during the floods.

In the past CWC prepared guidelines in 1873-74 for flood plain zoning which were approved by Central Flood Control Board. CWC also prepared a model draft and circulated it in the Ministry of Irrigation in 1975, to all the states for enacting legislature. However the response from states except Manipur has not been encouraging. Manipur enacted a legislation in 1978 which came into force in 1985.

Flood proofing measures, help greatly in the mitigation of distress and provide immediate relief to the population in flood prone areas. It is essentially a combination of structural change and emergency action, not involving any evacuation. The technique adopted consists of providing raised platforms for flood shelter for men and cattle and raising the public utility installation above flood levels.

In case of urban areas, certain measures that can be put into action as soon as a flood warning in received involve:

Installation of removable covers such as steel or aluminium bulk heads over doors and windows or other openings keeping stone counters on wheels, closing of sewer well, anchoring machinery, covering machinery with plastic sheet, seepage control etc.

Flood proofing also tends to encourage persistent human occupation of flood plains.

Out of the non-structural measures "flood forecasting and warning" is considered as one of the most important, reliable and cost effective methods. CWC organizes flood forecasting at 157 stations in the country, of which 132 are for water stage forecast and 25 for inflow forecast for certain major reservoirs. The Flood Meteorological Offices (FMO) also provide information regarding general meteorological situation, rainfall of last 24 hours for different regions and range of quantitative precipitation forecasts for various river basins to the respective flood forecasting centers of CWC. All the data is simultaneously transmitted to the circle headquarters supervising forecasting works for overall security, monitoring, analysis and compilation. The final forecasts are then transmitted to the administrative and engineering authorities of the state and other user agencies connected with flood protection and management work on telephone or by special messenger/ telegraph/ workers depending upon local factors like vulnerability of the area and availability of the communication facility etc.

The Master Plan 2000-2021 recommendations for river Jhelum were:

- All Doongas and houseboats be shifted upstream of Padshahibagh or downstream of Chattabal. The tourist oriented houseboats could be shifted to Dal or Nigeen area.
- Encroachments made on the banks of the river and all three Khuls- Kuta Khul, Soner Khul and Watel Khul be cleared en-masse.
- Development of river fronts will involve clearance of some sites for development of parks.
- To stop garbage dumping from Lal-Ded Hospital, incineration plant be installed there.
- Wherever possible Agriculture department should level the disused brick kiln sites on either side of cement bridge and use the same for growing vegetables.
- Shikara ghats be constructed at appropriate points, connected with pucca stairs to nearby roads.
- Inland water transport project be implemented as per project report. This would help to keep the water in turbulence, besides reduce pressure on road traffic and also serve as a tourit attraction. The vessels could also be used to ferry people in flood emergencies.

- While according permission for building construction on river and nallah fronts, no part of the building is protruded towards the river and nallah boundaries or over their embankments.
- Problem of river discharge on river Jhelum be solved as it has assumed serious dimensions. Over the past fifty years, river Jhelum and spill channel has heavily silted up. It was understood that the flood control problem was being entrusted by I&FC Deptt. to some consultancy firm, hence the problem of siltation, dredging, gradient, velocity etc. shall be dealt with. Some suggestions included to redesign flood absorption basin from Kandizal to Padshahibagh saving the railway line and Mahjoor nagar area. Beds of river Jhelum and spill channel be deepened to increase the discharge capacity and ensure the minimum draft required for mechanized water transport. Gradient of spill channel be increased up to the permissible limits. Weir at Chattabal be redesigned. Navigational channel of the lock gate at weir site be desilted to make it functional.

However the Master Plan 2000-2021 has also got entangled in the cobweb of red-tape like other similar vital issues which ultimately land us into a chaotic situation like the recent one. The process of reviewing the Master Plan is still in progress even in 2015.

C) TWO DAY SEMINAR ON FLOODS OF 2014 HELD AT LALIT GRAND PALACE HOTEL IN NOVEMBER-14

A two day national Seminar on "Retrospective and Prospective of 2014 Kashmir Floods for Building Flood Resilient Kashmir" was held at Srinagar from 15-16 November 2014. The Seminar was organized jointly by the Department of Earth Sciences, Kashmir University and

Centre for Dialogue and Reconciliation (CDR). The 2014 flood was triggered by the complex interplay of atmospheric disturbances that brought widespread and extreme rains all across the state. The Jhelum waters, that used to be the provider of life and sustenance, suddenly became a monstrously destructive force against the human life and the infrastructure that cohabit its backyards since millennia.

Any future flood strategies for Jhelum Basin shall benefit from our learning from this horrendous experience and the threadbare deliberations held at the two days National Seminar. The September 2014 floods were unprecedented in the flood history of Kashmir and got everyone concerned about the consequences of another such disaster if it recurred. The immediate steps to be taken are to develop a strategy for mitigating floods in the state and that requires realization across the region at the local and the national level. The aim of this symposium was to conduct deliberations with the select group of relevant people who have the expertise to recommend and formulate a long term action plan for flood disaster management and mitigation in the state of Jammu and Kashmir.

The experts from various central agencies (Central Water Commission-CWC, National Institute of Hydrology-NIH, National Geophysical Research Institute-NGRI, Central Groundwater Board-CGB, National Disaster Management Authority, NRSC/ISRO and National Green Tribunal), India Meteorological Department (IMD) and State Government agencies – Irrigation and Flood Control (IFC), Public Health Engineering (PHE), Rural Development, LAWDA, Srinagar Development Authority-SDA, IMPA, Agriculture Department, academia from Kashmir University, Indian Institute of Technology-Roorkee-IIT, National Institute of Technology -NIT-Srinagar, Jammu University, and various segments of the civil society, including experienced professionals, attended the Seminar.

Short-term and Urgent Recommendations:

The following recommendations made at the Seminar need to be taken up on priority immediately and could be accomplished in the shortest possible time to reduce the risk to the public and to property in Jhelum Basin from flooding.

01) Knowledge driven all-inclusive multidisciplinary flood planning needs to be initiated on priority by engaging technocrats with relevant expertise to develop insights into flooding mechanisms in the Jhelum Basin building on comprehensive existing studies.

02) Strengthening the flood infrastructure in the Jhelum Basin to cope up with the probability of next extreme flooding event of the magnitude observed in 2014. This includes the preparation of an integrated DPR for the construction of the alternate flood channel from Dogripora to Wullar, increasing the carrying capacity of the main Jhelum, dredging of the existing flood channel, dredging of the wetlands like Hokersar, Narkara, Nowgam Jheel, and Wullar lake, and strengthening of breached and weak embankments the broad plan of which is before the CWC.

03) The management of the water bodies/lakes and wetlands in the Jhelum Basin needs to be brought under one regulatory authority for their integrated management, being a single catchment area served by the same watershed.

04) The government, with the help of academia/research institutes, must consider undertaking a scoping study to assess the probability of flooding in immediate future based on the understanding to be developed from the interactions of Ground Water, surface water and the glacier-melt in the Jhelum Basin.

05) Urgently operationalising the Flood Early Warning System (FEWS) for Jhelum and Chenab.

06) The State Government must initiate on priority (with the help of leading academic institutions), to undertake transparent flood-zonation and flood vulnerability assessments of people and places at village level so that the flood risk reduction is integrated with developmental planning at village level in all District Development Plans.

07) Government consider assigning proposals for bringing the technical ingenuity of the Irrigation & Flood Control in operationalising of FEWS, basin wide IFM and flood scenario mapping. The identified scientific studies on various aspects of flooding identified above are required to be undertaken on priority by involving Universities, consultants and institutes both national and international.

Urgent Long-term Recommendations:

The following recommendations made at the Seminar need to be initiated immediately and might take a few years to complete for flood risk reduction in the Jhelum Basin.

11) There is an urgent need to institutionalize the disaster management in the state by setting up of a vibrant and structured State Disaster Management Authority with a clear mandate to build the capacity of the state to prepare for, protect against, respond to, recover from, and mitigate all types of hazards, the state is vulnerable to.

12) Strengthening of flood control infrastructure in 4 high gradient streams in the south Kashmir viz., Rambiara, Veshu, Romshi and Lidder, that enormously contribute to the discharge at Sangam using the available techniques, so that the flood peak and concentration time is appreciably delayed by staggering them in the watershed itself before their discharge into the Jhelum at Sangam.

13) Initiating a massive capacity building program for building public awareness and soliciting public involvement in flood risk reduction.

14) In order to arrest the siltation of the watercourses from the catchment, the participants recommended the massive reforestation of the Jhelum catchment under CAMPA, IWMP and other existing governmental schemes.

15) Structural and non-structural erosion control measures in the high gradient tributaries in the south Kashmir viz., Rambiara, Veshu, Romshi, Lidder, Bringi and Aripath.

16) Consolidation of the fragmented data and knowledge into a database so that it is available to everybody for use on understanding the hydrological and meteorological processes and phenomena in the state.

17) Strictly regulating mining of the riverbed keeping in view the river/channel morphology and other required hydrologic and geologic criteria.

18) The flood disaster preparedness at government and community levels need to be strengthened so that there is a well-rehearsed mechanism in place for quick response, despite all the adversities and limitations, to minimize the impacts of flooding on the people and property.

19) Revision of the existing land use policy and building codes is required and enforce strict implementation in order to minimize human and economic loss in the event of natural disaster.

20) Comprehensive community based Disaster risk reduction plans need to be prepared on priority.

Long-term Recommended Measures:

These long term recommended measures are essential for building the necessary flood control

Infrastructure in the basin so that in the eventuality of the next extreme flood event, the loss of life and property is reduced appreciably in the basin

e) Construction of the alternate flood channel from Dogripora to Wullar

f) Improving the drainage system in the urban areas of the Jhelum Basin including the restoration of natural drainages wherever possible

g) The government needs to initiate programs aimed at conservation and restoration of the degraded wetlands in the Jhelum Basin to enhance their flood mitigation, in selected cases even sewage treatment functionality.. Bring city and town planning in the state into consonance with the flood and earthquake vulnerability.

h) Structural and non-structural measures be initiated under the supervision of I&FC for erosion control in the central and north Kashmir part of the Jhelum basin.

There is encouraging news from todays GK regarding proposed construction of mini dams on tributaries to create a detention period for the flash floods as under:

JK plans 'mini dams' on Jhelum offshoots to avert 2014-like flood

'Feasibility study on, storage facility will off-load pressure on river'

The J&K Government is contemplating to build "mini dams" on tributaries of river Jhelum to offload pressure on the water body in extreme situations and avert the 2014-like floods in Kashmir in future.

The Flood Control Department Kashmir has started a feasibility exercise in this regard and is mulling to rope in expertise at international level for construction of water storage facilities, within the provisions of Indus Water Treaty, to address Kashmir's vulnerability to floods.

Chief Engineer of the Department JavedJaffer said the storage facilities would handle additional discharge from the tributaries during extreme weather conditions to ensure normal flow in Jhelum.

During last year's devastating deluge, Jhelum—against the present capacity of less than 30000 cusecs—recorded flow of over 1.20 lakh cusecs following continuous rainfall for almost a week, which ultimately saw water currents breaching embankments of the river and flooding the Valley.

A study by J&K's Department of Ecology, Environment and Remote Sensing, inCOLLABORATION ⌐ with the National Remote Sensing Center, has cautioned that parts of Kashmir would witness increase in the intensity of rainy days by 20 percent by 2030 and it could result in frequent flood like situations.

Though setting up of the dams would be a long drawn process and could take many years, the Department sees it as the "real solution" to the future flood threats in Kashmir.

A panel constituted by the Union Ministry of Water Resources to study the possible causes of the flood has also recommended setting up of the storage facilities.

At least 24 tributaries are the main feeding source of 150 miles long Jhelum, which originates in south Kashmir and snakes through Srinagar and north Kashmir before emptying into Wullar Lake in north Kashmir. Some of the tributaries drain from slope of PirPanjal range and join the river on left bank while some others flow from Himalayan range and join Jhelum on the right bank.

Jaffer said the Department would identify the tributaries and the potential spots for setting up storage facilities and then experts would be roped in for undertaking environmental impact and geological studies before the construction work is taken up.

He said the water storage facilities would also help the government to reduce dimensions of the proposed Rs 18000-crore new floodspill channel, proposed from Sangam in south Kashmir to Wullar, and estimated to carry additional 55000 cusecs of water.

The Government of India had sought a fresh Detailed Project Report from State government on the spill channel but more than a year after the flood struck Kashmir—killing more than 300 people and causing huge losses to housing and business sectors—there has been little progress on the report.

The Government has also been accused of going slow on increasing capacity of Jhelum to original 35000 cusecs and restoring capacity of existing flood spill channel to over 15000 cusecs.

Though Kashmir has a long history of floods, the Government has over the years allowed constructions, both in private and public sectors, along river Jhelum, and the encroachment of marshy lands, river bodies and lakes which has increased Kashmir's flood vulnerability.

A senior official of the Department said the dams would lessen the pressure on proposed flood spill channel and Wular Lake.

"The storage facilities will be set up strictly within the provisions of the Indus Water Treaty. We are also improving upon the DPR about the new flood channel before submitting it to Centre for funding," said the official.

The need for addressing Kashmir's vulnerability to flood has also grown after the Union Water Resource Ministry panel highlighted that bowl shape of Kashmir and "very mild slope of Jhelum" makes the Valley "highly susceptible" to flooding, at a time when the region is witnessing impact of climatic changes in the form of fast receding glaciers and frequent

10) Making Indian Engineers World-class

"Contribution of World-class Engineers"

World-Class Engineers are:
Solidly grounded in fundamentals of their discipline and are committed to lifelong learning.
Technically Broad : Conversant in multiple technical disciplines. They design solutions that span business functions such as finance, marketing, legal, and manufacturing.
Globally Engaged : Understand the worldwide nature of their profession and are sensitive to the speed required to keep pace in geographically and culturally diverse environments.
Ethical: Uphold the highest ethical standards. They readily identify, and carefully address, ethical issues that arise in their professional lives.
Innovative: Develop precise definitions of complex problems and formulate sustainable solutions by thinking creatively across technical, business, social, and environmental dimensions.
Excellent Collaborators: Seek optimal outcomes through collaboration and honor intellectual property rights of all partners. They work effectively within co-located and geographically dispersed teams.
Visionary Leaders Are courageous, customer-oriented leaders who develop visions that deliver successful results.

It is a fact that from the earliest times, the engineers, who remained the harbinger of development of any region, always strived for enrichment of their knowledge and skill to upgrade the quality of life and their performance. The pursuit for betterment is a continuous process. There is no end to development and engineering progress. The process of up gradation from one standard to the other, from one age to other, from 'under-developed' to 'developing', or from 'developing' to 'developed', is a continuous process, which is led by the engineers after taking into account the prevailing socio- politico-economic conditions of the particular country.

From the archival discoveries it is revealed that it has taken thousands of years for man to reach the present state.

The Prehistoric World: 100,000-40,000 years ago.

The first modern humans emerged in Africa 100,000 years ago. Over the next 50,000 years they colonized much of Asia and Australasia before expanding into Europe. New skills were acquired at different rates in different regions but the landmarks of development followed a similar pattern from simple stone blades to sophisticated iron jewellery. The different ages witnessed various stages of development such as:

Stone Age: Upper Plaeolithic - 40,000-10,000 years ago.

Early humans were already expert flint workers by the upper Plaeolithic period and weapons have been found at sites in Europe and the Near East. Typical features included:

- Stone spear heads, arrow heads and blades;
- Bone and ivory tools and weapons, (fish hooks, needles and spear throwers);
- Jewellery and clothing made of skins sewn using bone needles;
- The ceremonial burial of the dead;
- Cave art and statues.

Neolithic- from 12000 years ago.

The later Stone Age saw the development of farming which replaced hunter gathering as the primary mode of existence. By the end of the Neolithic, humans had learned to cultivate many crops: wheat and barley in the Near East, rice in China and potatoes in South America. Farming created surpluses, allowing population growth and establish permanent settlements. Other features of the period include:

- The domestication of the animals (by 6000 BC in China and Mesopotamia);
- New tools like axes to clear forests and bring new lands under cultivation, hoes, sickles and grindstones;
- The use of pottery to store grain;

- The construction of earliest villages and towns often surrounded by walls to coral livestock;
- Tombs built of stones.

The Metal Age:

Bronze Age: from 3000 BC:

- Copper and bronze tools and weapons (spearheads, arrowheads, chisels, saws);
- Practice of trade throughout Europe;
- Early mines and ore extraction methods;
- High standard for craftsmanship (jewellery, statues, decoration);
- Creation of stone alignments.

Iron Age: The Hitites of Anatolia made iron weapons between **2000 and 1200 BC**. Iron working spread to Greece in about 1000 BC. It had advantage over bronze as it gave sharper, harder wearing edge; no combination with other metal needed; supplies were plentiful; used for nails, tools, weapons, casting utensils, jewellery and also for religious articles. European Iron Age ended with Roman Empire. There was no Iron Age in Americas, where iron was introduced by European colonists.

Dawn of History: Civilization is closely linked to the rise of cities. Urban life emerged as agriculture started to support artisans, traders, government and organized religion as well as people living in the land. From about **3000 BC**, cities grew up on the banks of the Tigris and Euphrates rivers in Mesopotamia (Between the rivers), part of the 'Fertile Crescent'. They were independent city-states at first, then part of empires. At the same time Egypt grew in power, and the eastern Mediterranean became a crossroads for traders and empire-builders.

Ancient Greece: The essential characteristics of European culture and civilization were forged in Greece, which became dominant force in the Mediterranean for **400 years before Alexander the Great** briefly created one of the Largest Empires of the ancient world, spreading Greek (Hellenistic) culture to Egypt and deep into Asia.

Ancient Rome: Rome flourished for about **800 years**, developing a technically advanced and sophisticated society, not seen again in the Western world until the **16ᵗʰ century**. The early Roman state was a republic, ruled by a senate of leading citizens with elected magistrates or consuls. Despite frequent mismanagement, Rome sustained the empire for 400 years.

The Making of Europe: The collapse of Roman world left a mosaic of competing kingdoms in Europe. But most of the Germanic tribes were highly Romanized, had fought for Romans as mercenaries and had adopted their Christian religion. The changes they brought about were often more evolutionary than sudden. It was a time of turmoil, but out of the turmoil emerged new peoples and powers- and a new stage of European history.

Christianity: Within a few years after Christ, Jesus' message spread beyond the Jews and grew into a cult stretching across the Roman Empire. When the Empire collapsed, the Western church presented much to learning and traditions of Rome, eventually becoming the dominant force of the medieval world.

The rise of Islam: In AD 610 after a series of revelations the Prophet Muhammad (S.A.W.) founded a religion based on faith in a single God, clear social rules and the promise of afterlife. Arab conquests quickly spread Islam through south-west Asia, the Middle East and North Africa. Christian Europe was hostile to Islam but later benefited from the preservation of Greek culture, and the scientific and medical knowledge of Arab Muslims. Islam gave unprecedented impetus to the intellectual development of the human race and that early Muslims held high the torch of light and learning at a time when the whole world was immersed in ignorance and barbarity. Islam furthered the cause of science. Modern science owes its origin to Islam and modern progress is the outcome of the freedom of thought and spirit of enquiry proscribed for Muslims by the Holy Quran, and not a product of Christianity which for a long time relentlessly proscribed all free thinking and liberal reasoning and even scientific researches on original lines, and horribly persecuted all those who indulged in these. Muslims laid the foundation of Physical Sciences. Western Civilization is the direct offspring of Arab Civilization in Spain. The very Renaissance was brought about by the impact of Islamic culture and learning. All the knowledge, whether of Astronomy, Mathematics, Architecture, Physics, Medicine, History, Geography, Alchemy and Algebra, Modern Chemistry, Political Economy, Sociology, Zoology, Geology, Botany, Navigation, Agriculture, Irrigation, Gardening, Statistics, Chronology, Topography, even Aviation or Philosophy of which the Europeans later made themselves masters, originally derived from the Saracenic schools.

The Middle Ages: Around 1000 AD. Europe was divided among many monarchs and regional lords whose authority over their territories varied greatly. Trade expanded, towns grew and won autonomy, craftsmen formed guilds, and universities were founded. Writers such as Dante and Chaucer produced masterpieces, and massive cathedrals were built to assert belief in the power of the divine order.

India: There was a flourishing civilization in **Indus valley by 2500 BC**. Repeated invasions from Central Asia brought a succession of empires, influenced by first Hinduism and Buddhism and then by Islam. The last of these was Mughal Empire. But India's wealth and sophisticated economy continued to attract both trade and military invasion from the east as well as the west- most spectacularly of the British Raj.

Kashmir has the distinction of being the only place in the world that has a recorded history for the past about 5000 years. According to Rajtarangni kingship was established here right in 12th century BC itself.

China and Japan: For most of the world history, China was the richest and most powerful nation on earth. Until the 19th century, it remained almost self-sufficient, amassing huge national wealth by exporting silk, spices and (later) porcelain. Japan remained culturally in the shadow of its powerful neighbor for many centuries, but was equally insular and self-reliant.

Africa: The vast scale and the natural wealth of Africa are matched by a diversity and richness of culture. From the 1000 year kingdom of Meroe in southern Egypt to the fabulous wealth of the West African Gold Coast to the mysterious builders of great Zimbabwe, African people traded, worshipped and built empires across a vast continent. **Arabs** arrived from **7ᵗʰ century** and **Europeans** from the **15ᵗʰ**-first in search of trade, then as settlers, farmers and adventurers drawn by tales of minerals, gems and gold.

Ancient America: The people of ancient America developed distinctive civilizations in almost total isolation from the rest of the world. In Mexico, Central America and the Andes, farming peoples created complex urban societies centered on religious cults. Their cultures spread to the hunting and farming societies of North America. All these cultures were destroyed after the arrival of Europeans in 1942.

The Renaissance: In 14ᵗʰ century a new mood of enquiry stirred in Italy, and spread across Europe. Inspired by rediscovery of classical learning by Arabs, scholars and artists began to reappraise the world and it took **200 years** for the transition from medieval world to a modern one.

The age of exploration: In **15ᵗʰ century**, improvement in shipping and a demand for Far Eastern silks and spices led European navigators to explore new waters. The Portuguese followed Arabs and worked around Africa to India and beyond, while Columbus crossed the Atlantic. The whole world was now open to European exploration, trade and settlement.

This was followed by clash of faiths between Catholics and Protestants in **16ᵗʰ century**, in which lakhs of people were caught up in the struggle between the two faiths. Next the world witnessed the age of kings followed by European turmoil, creation of USA, the industrial revolution in 18ᵗʰ and 19ᵗʰ centuries, formation of new nations and empires in 19ᵗʰ century, world war I, Russian revolution, World war II, end of empire, The Cold war, The New world order in 20ᵗʰ century.

Landmarks of Civilization: These emerged in the Fertile Crescent after **10,000 BC**.

- **Cities**: Some of the oldest cities were found in the Middle East such as Jericho-8350 BC, Catal Hayuku in Anatolia –the largest city in the world **6250-5400 BC**.

- **Wheel:** It started off in Mesopotamia in **3500 BC** as a potter's tool and was used for vehicles after **3500 BC**.
- **Legal system:** Hamurabi (**1792-1750 BC**) king of Babylon codified the oldest known laws. The Jewish Torah dates from **4th century BC**.
- **Writing:** **Around 3300 BC** –the Sumerians developed one of the earliest writing system- a picture based script called cuneiform, impressed on clay tablets. In 1100 BC Phoenicians created a sound based alphabet later the basis of all modern European scripts.
- **Mathematics:** The number system of Mesopotamia gave us the 60 minute hour and 360 degree circle. The Arabic numerals with Indian zero was a great leap forward in this direction.
- **Monotheism:** Belief in a single all powerful God was a key feature of Judaism and later of both Christianity and Islam.

Scientific Thinkers: The search for a framework of knowledge about the world around began with the theorizing of ancient philosophers. By the 17th century, experimentation and observation were the preferred tools of deduction. In both approaches, progress has relied on a few exceptionally creative thinkers. Progress was made in the fields of mathematics, matter and energy, earth sciences, cosmology, life sciences and arts, photography, architecture, furniture, classical and popular music, dance, literature, drama, cinema, printing, newspapers, radio, television, fashion, food, games, road transport, trains, steamships, navigation, aviation, space travel, information technology, energy consumption, fossil fuels, nuclear power, renewable energy, mineral resources, electricity and magnetism, radioactivity, chemistry, archaeology, everyday inventions, telecommunications, computer technology, digital communications, the internet, modern medicine, civil engineering, age of armour, nuclear age etc. In all these fields engineers have played a key role. The pace of development in the past few decades has been much faster than ever before and the future poses more drastic challenges due to population explosion and limited available resources.

Thus Engineers are key figures in the material progress of the world as rightly portrayed in today's theme of the seminar.. A world-class engineer, regardless of the job he is engaged in, is always considered an asset to the nation and the society; as it is he who makes a reality of the potential value of science by translating scientific knowledge into tools, resources, energy, and labour to bring science into the service of the country.

It is a challenge to conclude about the class to which the engineers of India belong. In the diversified, heterogeneous nature of development in our country, the engineers have to work from construction of rural roads to manufacturing of spaceships to Mars. Both are equally important for accelerating the development of the country. There is no scope to undermine the contemporary skill and knowledge of the engineers of our country. It is a matter of pride that Indian engineers, whether working in the country or outside, are a force to reckon with globally.

There is a gap of development in the developed, developing and underdeveloped countries. However the lessons learnt from the experiences in the developed countries can be utilized in the development of the developing and the under developed countries. It was recently in the media that vertical expansion of the cities have not found favour in developed countries, and it was suggested to go in for the smart cities, which may also be suitable for the future development of Srinagar city Master Plan.

I have personally witnessed the excellent contribution of world class consultants and engineers besides the Indian engineers in the development of the modern cities in UAE and had an opportunity to attend the Infrastructure Arabia Summit conference alongside World eco-consult, 22-25 April 2012 Abu Dhabi National Exhibition Centre (ADNEC), extracts of which have been published by me in my book "Environment in Jammu & Kashmir" under the heading: "Building a sustainable framework for the Middle East compared to J&K State". Besides, American visitors expressed their opinion that the newly developed twin cities of Dubai and Abu Dhabi are far superior to the age-old developed American cities. One finds a marked difference between the huge steel sections of Indira Gandhi International airport and the aesthetically designed sections of Dubai airport. Similarly we find every road curve, foot path, signaling system, electric installations etc. strictly according to the prescribed engineering standards as against our constructions in J&K State, violating all norms without any consultancy or quality control. However the knowledge, skill, and wisdom of Indian engineers are no less than that of their counterparts from other so-called "advanced" countries. Due to the socio-politico-economic structure of our country, engineering is still very much labour-intensive. Unlike in other parts of the developed world, Indian engineers are quite capable of blending the modern mechanized systems with prevailing traditional human-oriented activities.

As rightly said, that it does not mean that the pursuit for self-enrichment by Indian engineers will not be perceived. India requires large numbers of qualified and competent engineers to address the numerous challenges faced in the developmental journey. To produce large numbers of competent engineering and technical personnel to take on the global challenges, India will need to complete the following activities to transform the curriculum for training and skill up gradation:

i) Generate awareness about the global nature of the profession, in-tune with growing challenges and opportunities – In this connection as seeing is believing, exposure of engineers to the problems could be achieved by arranging their tours to the developed and the developing countries.

ii) Develop a comprehensive understanding in the respective engineering discipline to tackle complex, real-world problems.--One need to keep abreast with the current challenges of engineering issues through modern information technology and enrolling as members of the national and international societies of engineers.

iii) Accept challenges and solve them with wisdom and shared knowledge -- Our engineers are equally competent to take challenges and find their solutions if given the opportunity.

iv) Acquire knowledge and expertise through lifelong education and continuous learning – In the present age of developed information technology this has become easier than before. Besides participation in seminars helps exchange of thoughts and knowledge.

v) Build familiarity in other engineering and scientific disciplines so that interdisciplinary solution approaches can be evolved.—Holding of interdisciplinary seminars, participation, interaction and exchange of thoughts can be of great help in this direction.

vi)   Pursue opportunities to apply skills in both traditional and non-traditional fields to address societal challenges -- Our engineers are quite competent to undertake this task provided given the opportunity.

vii)   Communicate and interact with other highly recognized international leaders in engineering, (again present facilities are far better for this job) and

viii)   Establish themselves as personalities with ethical and noble values—This is most important aspect for which moral education right from the school days needs to be stressed. In earlier days, there used to be taught to children the stories with moral endings like Shaikh Sadi's Karima Nami Haq, Gulistan, Bostan, Moulana Rumi's Masnavi and also Ikhlaq-i-Mohsini etc. that would remain inscribed in the child's mind all along his life and would help to build his character. This aspect has been ignored in the modern education system and that is why we are confronted nowadays with moral degradation around us.

It is rightly said that achieving excellence is a journey that needs considerable effort. It requires a transition from a reactive, compliance-based approach to a proactive, contributory and value-add mindset to create an environment of sustained operational progress. Over the long-term, it is hoped that the world-class engineers will create a set of approaches and best-practices that will improve tomorrow's world, create long-term value, and institutionalize business sustainability.

It is engineers who have contributed their bit in creation of world wonders in the past and the process is on with achievement of new discoveries in different branches of science and technology. Right from making of a needle to the creation of spaceship engineering is involved at every step. It is only when blood and sweat is put together that a wonder comes into being. As Dr Iqbal rightly said:

نغمہ ہے سودائے خام خون جگر کے بغیر
نقش ہیں ناتمام خون جگر کے بغیر

References:

01. Facts at your fingertips-Reader's Digest.
02. Islam's contribution to Science and Civilization- Maulvi Abdul Karim.
03. Wikipedia – the free encyclopedia

Seven Wonders of the Ancient World

The classic seven wonders were:

Great Pyramid of Giza, Hanging Gardens of Babylon, Statue of Zeus at Olympia, Temple of Artemis at Ephesus, Mausoleum at Halicarnassus, Colossus of Rhodes, Lighthouse of Alexandria-The only ancient world wonder that still exists is the Great Pyramid of Giza.

Lists from other eras:
In the 19th and early 20th centuries, some writers wrote their own lists with names such as Wonders of the Middle Ages, Seven Wonders of the Middle Ages, Seven Wonders of the Medieval Mind, and Architectural Wonders of the Middle Ages.
Stonehenge, Colosseum, Catacombs of Kom el Shoqafa, Great Wall of China, Porcelain Tower of Nanjing, Hagia Sophia, Leaning Tower of Pisa
Other sites sometimes included on such lists:
Taj Mahal, Cairo Citadel, Ely Cathedral, Cluny Abbey
Recent lists:Following in the tradition of the classical list, modern people and organizations have made their own lists of wonderful things ancient and modern. Some of the most notable lists are presented below.

American Society of Civil Engineers:In 1994, the American Society of Civil Engineers compiled a list of Seven Wonders of the Modern World, paying tribute to the "greatest civil engineering achievements of the 20th century":

Wonder	Date started	Date finished	Location
Channel Tunnel	December 1, 1987	May 6, 1994	Strait of Dover, between the United Kingdom and France
CN Tower	February 6, 1973	June 26, 1976, tallest freestanding structure in the world 1976–2007.	Toronto, Ontario, Canada
Empire State Building	January 22, 1930	May 1, 1931, Tallest structure in	New York, NY, U.S.

		the world 1931–1967. First building with 100+ stories.	
Golden Gate Bridge	January 5, 1933	May 27, 1937	Golden Gate Strait, north of San Francisco, California, U.S.
Itaipu Dam	January 1970	May 5, 1984	Paraná River, between Brazil and Paraguay
Delta Works/Zuiderzee Works	1920	May 10, 1997	Netherlands
Panama Canal	January 1, 1880	January 7, 1914	Isthmus of Panama

New7Wonders Foundation:

Wonder	Date of construction	Location
Great Wall of China	Since 7th century BC[16]	China
Petra	c. 100 BC	Jordan
Christ the Redeemer	Opened October 12, 1931	Brazil
Machu Picchu	c. AD 1450	Peru
Chichen Itza	c. AD 600	Mexico
Colosseum	Completed AD 80	Italy
Taj Mahal	Completed c. AD 1648	India
Great Pyramid of Giza (Honorary Candidate)	Completed c. 2560 BC	Egypt

Top 10 Engineering wonders of the modern world:

01. Pearl Bridge Japan – Longest suspension bridge-central span 6532 ft. – completed 1998

02. Millau Viaduct   -- Tallest cable stayed bridge – France- ht. 343 mts.- completed 2004

03. USS George H.W.Bush (CVN 77)- world's largest warship- 100,000 Mts.-comp—2009

04. North European offshore gas pipeline-Russia to Germany--- 1,222 kms. long 2011-2012

05. Beijing National Stadium China- world's largest steel structure - used in 2008 Olympics

06. Bailong Elevator China- world's highest and longest glass elevator-330 mts. High-2002

07. Palm Islands Dubai-world's biggest artificial islands- 1500 villas – on artificial beaches

08. Euro Tunnel- England to France- underwater -31 miles long, 23 of which is in sea.

09. Three Gorges Dam China- Hydroelectric dam on Yangtze River-22,500 MW- com.2008

10. Pan-STARRS-for Panaromic Survey & Rapid Response System- see galaxy ever better.

It is only when blood and sweat is put together that a wonder comes into being. As Dr. Iqbal rightly said:

نقش ہیں نا تمام خون جگر کے بغیر

نغمہ ہے سودائے خام خون جگر کے بغیر

11) Seven Billion Dreams. One Planet. Consume with Care.

The well-being of humanity, the environment, and the functioning of the economy, ultimately depend upon the responsible management of the planet's natural resources. Evidence is building that people are consuming far more natural resources than what the planet can sustainably provide. Many of the Earth's ecosystems are nearing critical tipping points of depletion or irreversible change, pushed by high population growth and economic development. By 2050, if current consumption and production patterns remain the same and with a rising population expected to reach 9.6 billion, we will need three planets to sustain our ways of living and consumption. The WED theme this year is therefore **"Seven Billion Dreams. One Planet. Consume with Care."** Living within planetary boundaries is the most promising strategy for ensuring a healthy future. Human prosperity need not cost the earth. Living sustainably is about doing more and better with less. It is about knowing that rising rates of natural resource use and the environmental impacts that occur are not a necessary by-product of economic growth.

Announcing World Environment Day 2015, UN Under-Secretary-General and UNEP Executive Director Achim Steiner said, **"While industrialized countries account for the bulk of the world's resource consumption, unsustainable consumption patterns are becoming more prevalent worldwide, with 3 billion middle class consumers expected to be added to the global population by 2030 - many of them from emerging economies."**

"Food production is one of the most obvious examples of unsustainable consumption patterns, with 1.3 billion tonnes of food being wasted every year, while almost 1 billion people go undernourished," he added. "This is an issue that UNEP is helping to address with partners like the UN Food and Agriculture Organization (FAO) through our joint campaign against food waste, **Think.Eat.Save**. We are glad the Expo's theme also focuses on sustainable food systems."

"World Environment Day provides us with an important opportunity to identify solutions for re-engineering our consumer culture to create a sustainable society in which everyone has enough to live well while staying within the planet's regenerative capacity. It is time to look seriously at what our appetite-for-more is costing the planet, our health, our future, and the future of our children," he said.

Feeding the Planet- Energy for life

About 7 billion people are alive today. Every second three more are added to the total, a growth of more than 10,000 an hour, over 80 million in the space of a year. The world population has almost tripled since 1950.

The world population (the total number of living humans on Earth) was 7.244 billion as of July 2014 according to the medium fertility estimate by the United Nations Department of Economic and Social Affairs, Population Division and it was projected to reach 7.325 billion in July 2 0 1 5 .

World Population: Past, Present, and Future

At the dawn of agriculture, about 8000 B.C., the population of the world was approximately 5 million. Over the 8,000-year period up to 1 A.D. it grew to 200 million (some estimate 300 million or even 600, suggesting how imprecise population estimates of early historical periods can be), with a growth rate of under 0.05% per year.

A tremendous change occurred with the industrial revolution: whereas it had taken all of human history until around 1800 for world population to reach one billion, the second billion was achieved in only 130 years (1930), the third billion in less than 30 years (1959), the fourth billion in 15 years (1974), and the fifth billion in only 13 years (1987).

Even though the rate at which it is growing has slowed down from its peak 2.4% a year in 1965, it has risen to over 6 billion by the turn of the century. The maximum expansion has been fueled not by an increasing birth rate but by a gradual extension of life expectancy and by a huge reduction in the number of children who die when young. More than half the people now living are under 25. In Africa almost half are under 14.

In Europe, Japan and North America, however the number of births has already dropped so as to be almost in balance with the number of deaths and the population there is now virtually stable. For the world as a whole, the balance will not come, according to UN predictions, until about the year 2110 when there might be 10.5 billion people striving for living space.

- During the 20th century alone, the population in the world has grown from 1.65 billion to 6 billion.
- In 1970, there were roughly half as many people in the world as there are now.
- Because of declining growth rates, it will now take over 200 years to double again.

World Population Milestones

8 Billion (2024)
According to the most recent United Nations estimates, the human population of the world is expected to reach **8 billion people in the spring of 2024**.

7 Billion (2011)
According to the United Nations, world population reached **7 Billion on October 31, 2011**.

The US Census Bureau made a lower estimate, for which the 7 billion mark was only reached on March 12, 2012.

6 Billion (1999)
According to the United Nations, the **6 billion figure was reached on October 12, 1999** (celebrated as the Day of 6 Billion). According to the U.S. Census Bureau instead, the six billion milestone was reached on July 22, 1999, at about

3:49 AM GMT. Yet, according to the U.S. Census web site, the date and time of when 6 billion was reached will probably change because the already uncertain estimates are constantly being updated.

Previous Milestones

- **5 Billion**: 1987
- **4 Billion**: 1974
- **3 Billion**: 1960
- **2 Billion**: 1927
- **1 Billion**: 1804

Growth Rate

Population in the world is currently growing at a rate of around **1.14%** per year. The average population change is currently estimated at around 80 million per year.

Annual growth rate reached its peak in the late 1960s, when it was at 2% and above. The rate of increase has therefore almost halved since its peak of 2.19 percent, which was reached in 1963.

The annual growth rate is currently declining and is projected to continue to decline in the coming years. Currently, it is estimated that it will become less than 1% by 2020 and less than 0.5% by 2050.

This means that world population will continue to grow in the 21st century, but at a slower rate compared to the recent past. World population has doubled (100% increase) in 40 years from 1959 (3 billion) to 1999 (6 billion). It is now estimated that it will take a further 43 years to increase by another 50%, to become 9 billion by 2042.

The latest United Nations projections indicate that world population will nearly stabilize at just above 10 billion persons after 2062.

Food:

Food is mankind's raw energy source-the fuel that fires the human boiler- and maintaining the supplies constitutes man's biggest single concern.

Agricultural efficiency has increased at a staggering pace: in 1980 the world's farms produced twice as much food as they did in 1950. As a result the earth today grows enough food to support its population, with plenty to spare. But the pattern of production is uneven and many areas still go short. Although the world produces enough food for everyone to receive an adequate diet, yet food shortages and famines still occur. The main causes include overpopulation, particularly in India and China, draught, which has repeatedly occurred in Sub-Saharan Africa, and war. Famines induced by human conflict have provoked international action to alleviate hunger and improvements in global communication.

Thousands of different types of plants are consumed by man, but just three- wheat, corn and rice – account for about half of the world's harvest.

By no means every corner of the planet's surface can be exploited for crop farming, however for a combination of three basic factors- sunshine, moisture and soil- determine where the global harvest can be gathered in. At present only 11 % of the earth's surface is farmed for crops, while a further 20 % is thought to be cultivable.

Beyond the Green Revolution:

Hunger became a global issue in the wake of first and second World Wars. International armed conflict disrupted food supplies highlighting the need for an agency to assure that every country could meet its food requirements. In 1945 the UN FAO was established with the intention of raising levels of food production and nutrition worldwide.

An acceleration in population growth after the Second World War reached its peak in the early 1960's. In 1963 the FAO launched the Green Revolution, aiming to provide enough food to cater for future population expansion. They developed higher yielding varieties (HYVS) of cereals such as rice, wheat and corn through selective breeding. HUYS produces three crops annually on the same land. Farmers also encouraged using high levels of fertilizers and pesticides to improve yields.

By the 1980's, wheat and rice yields had increased dramatically; some developing countries produced a surplus for the first time. The increased income led to mechanization of more farms, further increasing yields. Agrochemical companies making fertilizers and pesticides grew into large business. Worldwide food production has doubled in the last 50 years.

Thus one major factor in the recent boom in food production has been the development of new high-yielding strains of wheat, corn and rice. Cultivated with modern fertilizers, these grains have generated what is known as green revolution.

The disadvantages: Overall the wealthier farmers benefitted from the initiative more than the poorer ones, who were unable to afford the new HYV seeds, fertilizers and pesticides. Mechanization led to unemployment, and repeated cropping damaged the soil. Chemical fertilizers and pesticides were found to be toxic to workers and caused pollution.

Bumper harvests have been the result. But the techniques modeled on the practices of US agriculture, have had their critics too. The chemicals required by high yield strains are derived chiefly from fossil fuels, which have become increasingly costly since the oil crisis of 1973. Mechanized farming too drains out energy resources- about 2 gallons of gasoline are required to produce an acre of corn in USA. Such farming tends to benefit the large farmer with capital invest at the expense of small farmer.

In coming decade the world may well see some adjustment toward the kind of organic farming practiced in China. Here crop waste is recycled to provide fertilizer, so that less synthetic matter is required. In addition mixed cropping is practiced: grains are planted with legumes, such as peas and beans, which produce their own nitrogen for fertilizer through the bacteria in their roots. One legume – the Soy bean is already a post world war II success story in the developing world. It is grown increasingly for its high protein content and adaptability and the oil is used for making paints and chemicals, as well as margarines and cooking oils.

Some authorities believe the answer to world food shortages now lies in a new revolution based on genetically modified (GM) crops. Perhaps the greatest hopes for feeding future generations lie in the plant breeding and genetics. Resistance to pests and diseases can be bred into the crops so that spraying with hazardous chemicals becomes increasingly obsolete. Strains may be developed to cope with the harsh climates of desert and tundra. Modern techniques of gene transfer offer possibilities for cultivating radically improved species. Tens of thousands of potentially edible species have been identified and it may yet prove possible to carpet the world's most barren wastes with new forms of nutritious vegetation.

Global calories consumption: According to UN, an average adult should consume a minimum of 2400 kcal per day to lead a healthy, active life. Thus who are more active or live in cold climates, require more calories than those who are less active, or live in tropics. In countries, where the average daily consumption is 200 kcal or less (roughly equivalent to 2.2 kgs of potatoes or 1.4 kgs of rice). The majority of the population is malnourished. Some 800 million people in the developing world do not get enough to eat. In the developed world, about 34 million people have poor diets and unreliable food supplies.

Food Chains:

Sunshine: The sun is at the start of food chain. And since scientists believe that sun has another 5 billion years of life ahead, the first of agriculture's major needs appears to be well catered for. Plants convert solar energy into food through photosynthesis. In general therefore, the more sunshine there is, the higher crop yield. Even the most efficient crop cover, however cannot convert more than about 3 % of incoming solar energy into the chemical energy of plant growth, or biomass. Only a fraction of this biomass will be edible food. The levels of solar energy in different parts of world are:- Above average, Average and Below average.

Water: The sunnier the climate is, the more water plants need- for almost as much water is lost from a field of corn as would be evaporated from a lake of similar size. Although there is no shortage of water on a global scale, its distribution is uneven. Farmers adjust to water deficits by irrigation. In China 46 % of agricultural land is irrigated. In North America and Europe, where most areas receive sufficient rainfall, about 10 % of the land now benefits from some supplementary watering. The water budget in the globe ranges from large surplus to small surplus to small deficit and large deficit.

Soil: Crops need more than just sunshine and water. They also require soils rich in nutrients, such as nitrogen, potassium and phosphorous. Naturally fertile soils account only as small proportion of the earth's surface and are not evenly distributed among the continents. Even these soils need to be rested or left fallow, unless the farmer can replace what he has removed in the crop. Soils that are not naturally fertile can be enhanced by the use of fertilizers.

How the planet provides:

Climate and environment are the world's great chefs, giving Mexico its torillas, Greece its goats milk cheese, China its pork spareribs, and Japan its seafood dishes. And it is regional variations in these two factors that strongly influence what is raised where.

As stated the world's three main cereals are wheat, corn and rice, each of which has its special needs. Wheat is a crop of temperate prairies and will tolerate very cold winters. Corn is vulnerable to frost and it is therefore confined to a warmer climate band. And rice favors the special combination of warmth and copious rainfall that is found especially in monsoon zones.

156

Grain constitutes about half of the world's food production by weight, but similar factors associate other crops with particular environment for example, grapes with Mediterranean climates and the potato with dull, cloudy skies and clammy soils. Feast and Famine: If the global harvest were to be shared equally, each person could receive 2.3 kgs of food per day. Hunger need never be with us.

The reason why famines still take their terrible toll has more to do with the complexities of politics, economics, storage, and distribution other than physical capacity of the earth itself. The planet is fertile. Science has opened up new possibilities. And in the opinion of many experts, the age old scourge of hunger could with global cooperation have been eradicated by the end of last century.

To meet the future needs, we can colonize the world's inhospitable areas. The earth's total cultivable land is some 7.9 billion acres (3.2 billion hectares) of which less than half is currently being farmed. Although the remainder may be harsh or inaccessible terrain, we have the means to drain swamps, plant hill sides, and bring deserts into bloom.

One short term response to starvation in the third world is to transport surplus food from where it is stock piled and where it is needed. The biggest grain exporters are the United States, Canada, Australia and Argentina. Thanks to the Green Revolution, India, Thailand, Burma and Suriname can now be added to the list of smaller net exporters. Many others for example Mexico and Russia would be net grain exporters but for the demands of livestock which now consume more grain than grass.

In the long term, however this does nothing to help farmers in poor countries to produce more. Pouring cheap food in the third world can lower the prices there so much that local farmers are put out of business. Except in emergencies perhaps what poor countries need most is appropriate technology, transport facilities, education, and better administration.

Fruits of the earth: The crops that feed the world fruits have been cultivated as food for about 10,000 years. Until recently, however, the diet of a particular region was more or less based on what grew under local climate and soil conditions. In the 20th century, science, along with a revolution in communication and transport, has created a lucrative global market in fruit and vegetables. Almost anything is now available anywhere at any time – for a price.

Breeding and Rearing: Livestock provides mankind with food, clothing and muscle power. Scientific breeding and advances in veterinary medicine are now helping to create more profitable animals, but there can be an unforeseen price to pay for this, in the form of poor quality meat, and the spread of deadly disease across the species.

There are vast expanses of desert and bleak uplands whose lean and rocky soils support little more than coarse grasses. Since human stomach cannot digest grass, it is the livestock here- in particular the sheep and the goats – that act as our food converters, yielding meat, milk and cheese.

Cattle can be raised for the world's overpopulated regions and due to their vulnerability to the tsetse fly, are especially scarce in the humid tropics. China is the main producer of pork, yielding nearly 40 % of the global total.

Harvesting the seas: Fishing has grown from supplying local needs into a major commercial enterprise. At the end of 20th century it was estimated that around 5 million people worldwide made a living from fishing, and for many countries, such as China, trade in fish products is vital to their economy.

In all major fisheries of the world catch sizes are increasing faster than breeding is capable of replenishing stocks. An outright ban has been placed on the fishing of endangered species in some areas. With effective management of depleted stocks, it was hoped that there will be an increase in world annual production to 144 million tonnes by 2010; without it there would be a shortfall of 20 million tonnes.

Methods of fishing: Modern fishing fleets use sophisticated equipment for locating and hauling in fish. Aerial surveillance along with computer-controlled, satellite and sonar tracking devices, and enormous nets ensure that catches are large.

Fish Farming: 20 % of the fish we eat comes from aquaculture, in intensive cultivation of fish. Fish eggs are placed in warm water tanks until they hatch into fry, and then reared in fresh water or sea water tanks or cages.

Asia has a long tradition of crop farming, and now produces 90 % of the world's output. Salmon and trout fisheries, which originated in Norway and Scotland, also flourish today in Chile and Canada, both major producers for the European market. The tilapia and perch like fish is being successfully cultivated in parts of Africa.

Fish like all other foodstuffs, display preferences for habitat. Cod favors the cold waters of the North Atlantic, while tuna prefers warmer seas; flat fish, such as halibut, feed on the seabed,- while herring cruises close to the surface. The principal fishing grounds are all in coastal zones where nutrients, leached from the land, mix with rich sediment that is swept up from the sea floor by ocean currents and offshore winds. These waters comprise our teaming marine meadow lands, thick with tiny plankton supporting larger organisms that are, in turn consumed by shoaling fish.

In total the earth's fishing fleets bring in some 68 million tons a year. Japan with its intricate network of islands has an ancient fishing tradition and remains the largest single harvester of the sea.

Water for Life:

Water is a limited resource, which needs to be carefully managed. Its natural abundance in a region, and how it is collected, stored and distributed, has a major impact on a country's economy, determining what crops can be grown, and whether there is sufficient to meet domestic and industrial demands. The establishment of the first civilizations in the Middle East was due to the inspired use of Nile flood water for irrigation.

The water we use: If water is to be available on demand all year round, it needs to be collected and stored. How this is done varies around the world, according to climate and geography.

The water in most rivers and lakes is clean enough to support wild life, but before it flows out of the tap it must be made safe for human consumption. This is achieved at a water treatment plant in a series of steps. But in spite of the fact that Kashmir Valley was supposed to have purest form of water, it is unfortunate that we have to import bottled water from outside the State.

The Aral Sea was once the world's fourth largest lake. But from the 1960's on, the rivers feeding it were diverted to irrigate the cotton fields of Kazakhstan and Uzbekistan. The sea began to recede, and its dwindling waters were irredeemably polluted with pesticides and other agrochemicals. The Aral is now an ecological disaster zone and as predicted it was supposed to disappear by 2015, leaving a poisonous desert in its place.

How we use water?: More than 90 % of world water consumption goes to agriculture. Domestic use accounts for less than 3 %, with only a little more being consumed by industry. The major industrial use of water is for cooling nuclear and other thermal power plants, and for turning turbines in hydroelectric plants. Other heavy industrial users are the chemical, oil, paper and machinery manufacturing sectors.

Land Irrigation: UNESCO estimates that nearly half the world's crop production, in terms of value, comes from irrigated land. Without irrigation many nations would find it impossible to feed their population or develop their economics. In China and India the high yield of rice is totally dependent upon controlled floods, which irrigate the river plains in the dry season. Egypt would be as fertile as the Sahara desert without the heavy monsoon rains from the East African highlands which flood the River Nile. The water is stored, via the Aswan High Dam, in the lake Nasser reservoir. The thriving fruit farms are dependent on water from the Colorado River via 390 km long aqueduct. The Desalination is the other process used by some Arab countries to boost their fresh water supply. Artesian wells also serve as a source of fresh water at some places.

The Energy resources:

Energy Consumption: Fueling the world-using natural resources to drive the global economy- The world's most developed countries are its most voracious consumers of energy. Every year the USA consumes energy in all its forms equivalent to around 8 tonnes of oil per head of population: its poorer neighbor Mexico consumes the equivalent of just 1.5 tonnes per head. Most of this energy is created by burning non-renewable resources such as oil, coal- how long these will lat depends on the speed of industrialization in currently underdeveloped countries, and on global efforts to conserve energy by using it more efficiently.

Measuring world energy consumption: Global energy consumption is measured in tones of oil equivalent, which includes all forms of energy from fossil fuels to alternative resources such as nuclear, hydroelectric, geothermal, wind and solar power. In 1998, the world consumed energy equivalent to more than 9.5 billion tones of oil- each person consumed an average more than 1.6 tonnes of energy.

Fossil fuels: Almost 80 % of the energy consumed globally is produced by burning fossil fuels- coal, oil (petroleum) and natural gas- the remains of living organisms that have been buried in the earth for millions of years. Fossil fuels are the cheapest and most effective way of energy, but resources are finite and are steadily being used up.

Alternative renewable resources: such as solar and water power will eventually have to replace fossil fuels as the world's major energy resource. I have seen Masdar City coming up in Abu Dhabi that is totally designed to use solar energy for all purposes like lighting, warming water, air conditioning, transportation etc. and no fossil fuels shall be used there. They are planning ahead before fossil fuels run out.

Carbon dioxide levels: Burning fossil fuels releases carbon dioxide, which contributes to global warming, by trapping infrared radiation in the atmosphere known as "the green house effect". China, the Middla East and the former USSR produce the largest amounts of carbon dioxide in relation to the amount of energy they create.

Oil: About 95 % of the world's oil has been produced by 5 % of its oil fields. Two thirds of the largest fields have been found in the Middle East. Scientists estimate that reserves will run out before 2060.

Natural Gas: Russia and Middle East originally contained the world's largest natural gas reserves. Only 14 % of global reserves have been used up, but it is estimated that remaining reserves are likely to run out before 2115.

Coal: Coal reserves exist in every continent, including Antarctica, but technology and economics will only allow the recovery of 7 %. Estimates of when reserves will run out range from 2250 to around 3400.

Nuclear Power: Nuclear power is generated by the fusion, or splitting apart of atoms of uranium or plutonium. The process releases huge amount of energy using small amount of raw material: the fusion of 1 kg of uranium releases as much energy as burning 2000 tonnes of coal or 8000 barrels of oil.- But this process too has its pros and cons like use of less raw material and non release of uncontrolled emission into the atmosphere unlike fossil fuels. However nuclear power stations are expensive to build and public concern has led to protests over the storage of highly radioactive waste and the danger it poses to human health, and the possibility of appropriation for the unlicensed manufacture of nuclear weapons.

Renewable Energy:

The unrelenting global demand for energy and the knowledge that fossil fuel will not last forever, has led to a hunt for renewable resources. The use of hydroelectricity is well established. In countries such as Norway and Brazil, it accounts for more than 90 % of domestic electricity generation. The oil crisis of 1970's created renewed interest in wind power, a field now led by Germany, the USA, Denmark and India. Tidal and wave power and wind power is also being developed at suitable places.

By the end of 20th century, several viable alternatives to fossil fuels have emerged. Solar power heats water in more than a million homes in Greece. Iceland capitalizes on its natural general resources to heat 85 % of its houses. Biomass energy produced by burning or chemical process of organic matter, provides 15 %*- of domestic power in Scandinavia and is the main energy source for millions in China and India.

Mineral resources: More than 2500 minerals have been identified. Their widespread occurrence and durability made them ideal for trading in the ancient world – bars of metal were exchanged for goods in Egypt as early as the 4th millennium BC. Today mineral and metals – even precious substances such as diamond and silver – are more vital to the global economy for their broad industrial applications than as a medium of exchange.

Precious metals and minerals: Scattered deposits of diamond, gold and silver were discovered in river beds in ancient times. The Romans began mining gold and silver in Spain in the first century BC. Platinum mining began after the first deposits were discovered in Colombia in the 16th century. Diamond mining began with the first discovery of rock-bound specimens in South Africa in 1870.

Non metals like fluorspar, phosphate, sulphur, potash were discovered in 12th century and put to industrial use. Similarly metals like aluminum, chromium, copper, iron, lead, magnesium, tin, titanium, tungsten, uranium, zinc, starting from 5000 BC, were found useful for use in machinery and in the production of electricity and nuclear power.

Sustenance in Earth for Men and All who search or Ask:

Allah has blessed the hills and the earth with sustenance in measured quantities for all who search and ask for it and on which all kinds of life depend and derive benefits to sustain themselves (41:10). Allah has provided all kinds of natural resources in measured quantities in the hills and in other parts of the earth for man and for all other living creatures i.e. sustenance from forests and deserts, plains, hills, mountains, rivers and seas etc. which yield all kinds of produce for sustenance i.e. fodder, food, shelter and mineral wealth, from which man and other living creatures like animals, birds, insects, aquatic life and all known and unknown wild life derive benefits.

The Holy Quran says:

وجعل فيها رواسى من فوقها وبٰرک فيها وقدر فيها اقواتها فى اربعته ايام سوآ للسآ يٰلين ة

"He set on the (earth), Mountains standing firm, High above it, And bestowed blessings on The earth, and measured therein All things to give them Nourishment in due proportion, in four Days, in accordance With (the needs of) Those who seek (sustenance)" (41:10)

وارزقنا وانت خيرالرٰزقين ة

"And provide for our sustenance, For Thou art the best Sustainers (of our needs)" (5:114)

Aameen!

12) KNOWLEDGE-DRIVEN ERA AND THE DAWN OF KNOWLEDGE PORTAL TECHNOLOGIES

Today we are living in an age of knowledge explosion. What we studied at Graduation level half a century back, is being taught now at Primary level and the school going children are feeling to be under stress. So is the case with the grown-ups; it has become difficult to keep pace with latest developments of knowledge as every hour some new discovery/ invention takes place. In earlier days the pace of development was much slower and it would take a long time for the news/technology to reach other places. Acquisition of knowledge has been human endeavor right from the beginning. According to a Hadith: "One has to learn knowledge right from cradle to the grave." From the following details, we can have an idea of the history of the development of knowledge right from the earliest times, when each era has been a *knowledge era* according to its own limitations:

The Dawn of History: Around 1-2 million years ago several humanoid species existed according to fossil finds. For 200,000 years, the dominant human species in Europe and Asia were Neanderthals, who were driven to extinction around 30,000 years ago, by new arrivals from Africa-Homo Sapiens.

The Neanderthals appeared in Europe about 250,000 years ago-the name comes from the Neander Valley near Dusseldrof Germany, where remains were first found in 1856. There is ample evidence that the Neanderthals were cultural beings. Skillfully wrought stone tools and jewellery has been found. Graves showed that they buried their dead with some ceremony. They also used fire - vital for survival in the cold climate of the period. About 30,000 years ago they were outclassed by the more adaptable Homo-sapiens also known as 'Cro-Magnon man', named after the place in Dordogne France, where they were first found.

Mitochondrial Eve: In 1986, researchers at the University of California concluded that all humans were descended from a single woman who lived in Africa some 200,000 years ago. They based this on analysis of DNA taken from the mitochondrial specific parts of the human cell. This DNA differs from DNA in the cell nucleus and it passes only through the female line. It mutates at a very rapid but steady rate. It appears that her lineage has survived to present day.

Various Ages:

The Stone Age -(Upper Paleolithic)- 45,000 – 10,000 years ago

Early humans were already expert flint workers by the Upper Paleolithic period. More than 100 distinct tools and weapons have been found at sites in Europe and the Near East. Typical features of Upper Paleolithic cultures included:

Stone spearheads, arrowheads and blades

Bone and ivory tools and weapons (fishhooks, needles and spear throwers)

Jewellery and clothing made of skins sewn using bone needles.

The ceremonial burial of the dead.

Cave art and statues.

The Stone Age –Neolithic- from 12000 years ago:

The later Stone Age saw the development of farming which replaced hunter gathering as the primary mode of existence. By the end of Neolithic, humans had learned to cultivate many crops, wheat barley in the Near East, maize in the Central America, rice in China and potatoes in South America. Farming created surpluses, allowing populations to grow and establish permanent settlements. Other features of Neolithic included:

- The domestication of animals (by 6000 BC in China and Mesopotomia)
- A new tool-for example, axes to clear forests and bring new land under cultivation, hoes, sickles and grindstones.
- The use of pottery to store grain.
- The construction of the earliest villages and towns, other surrounded by walls to corral livestock (Jericho and Catal Huyuk)
- Tombs built of stone.

The Metal Ages:

a) *The Bronze Age –from 3000 BC*

The first experiments were made in Iran and Turkey some 9000 years ago. Copper and gold were the first metals to be used for tools and weapons, followed by bronze (an alloy of copper and tin). The Bronze Age featured:

- Copper and bronze tools and weapons (spearheads, arrowheads, chisels, saws).
- The practice of trade throughout Europe.
- Early mines and ore extraction methods.
- High standards of craftsmanship, jewellery, statues and decorations.

b) *The Iron Age: from 1200 BC*

Iron was first used long before the Iron Age. The Hittites of Anatolia made iron weapons between 2000 and 1200 BC. Iron working spread in Greece in about 1000 BC, and to Northern Europe, Asia and Africa by about 750 BC. It was brought to Britain by the Celts- members of Iron Age culture originating in the Austrian Alps. Iron had three advantages over bronze. It gives a sharper, harder wearing edge, it did not need to be combined with another metal, and supplies were plentiful. It was used for nails, tools, weapons, cooking utensils, jewellery and also for religious articles. The European Iron Age is conveniently said to end with the spread of Roman Empire. There was no iron age in Americas, where iron was introduced by European colonists.

Civilization is closely linked to the development of cities. Urban life emerged as agriculture started to support artisans, traders, government and organized religion as well as people living in the land. From about 3000 BC, cities grew on the banks of the Tigris and Euphrates rivers in Mesopotamia ('Between the Rivers') part of the 'Fertile Crescent.' They were independent city states at first, then part of empires. At the same time Egypt grew in power, and the eastern Mediterranean became a crossroad for traders and empire-builders.

Land marks of civilization: Many of the major development that we associate with Western civilization first emerged in the Fertile Crescent after 10,000 BC.

Cities: Some of the world's oldest cities are found in the Middle East, such as Jericho founded 8,350 BC, Catal Huyuk in Anatolia Turkey was the largest city in the world. It flourished 6,250-5,400 BC.

Wheel: The wheel started off in Mesopotamia 3,500 BC as a potter's tool. It was used for vehicles after 3,200 BC.

Legal Systems: Hammurabi (1792-1750 BC), king of Babylon, codified the oldest known laws. The Jewish Torah dates from the 4th century BC.

Writing: Around 3,300 BC, the Sumerians developed one of the earliest writing systems, a picture based script called Cuneiform, impressed in clay tablets. In about 1100 BC the Phoenicians created a sound based alphabet, later the basis of all European scripts.

Astronomy: The city of Ur was the birthplace of astronomy. By 1000 BC the Babylonians were predicting lunar eclipses and tracking planets.

Mathematics: The number system of Mesopotamia gave us the 60 minute hour and 360 degree circle.

Monotheism: Belief in a single all-powerful God was a key feature of Judaism, and later of both Christianity and Islam.

Important landmarks in the History of the world:

The Prehistoric World: 100,000 years ago
Dawn of History: 10,000 BC - 323 BC
Ancient Egypt: 5000 BC - 30 BC
Ancient Greece: 2000 BC – 146 BC
Judaism: 1200 BC – 70 AD
Ancient Rome: 753 BC – 406 AD
The making of Europe: 402 AD – 1066 AD
Christianity: 4-6 BC – 1453 AD
Islam: 570 AD – 1492 AD
The middle Ages: 1000 AD – 1485 AD
India: 5000 BC – 1857 AD
China & Japan: 6000 BC – 1905 AD
Africa: 7000 BC – 1914 AD
Ancient America: 5000 BC – 1890 AD

The Age of Exploration: 1492 AD – 1779 AD
The Renaissance: 1305 AD – 1633 AD
Clash of faiths (Christianity): 1415 AD – 1633 AD
Age of kings: 1643 AD – 1772 AD
Europe in turmoil: 1775 AD – 1815 AD
Creation of United States: 1607 AD – 1890 AD
The Industrial Revolution: 1701 AD – 1913 AD
New Nations and Empires: 1783 AD – 1901 AD
First World War: 1914 AD – 1918 AD
Russian Revolution: 1861 AD – 1924 AD
Second World War: 1939 AD – 1945 AD
End of Empire: 1940 AD – 1990 AD
The Cold War: 1945 AD – 1990 AD
New World Order: 1980 – to-date.

Adler's classification of knowledge:

All subjects fit under or under a combination of *constructs*, *science*, *engineering*, or *humanities*. Engineering overlaps with the other subjects. Mortimer J. Adler classified knowledge into six divisions: Logic, Mathematics, Science, History and the Humanities, Philosophy, and Preservation of Knowledge. The structure below is based on Adler's classification.

Constructs
Math
Logic
Philosophy (interdisciplinary)
Communication (interdisciplinary)
Engineering (interdisciplinary)
Science (Natural science)
Astronomy
Biology
Medicine (interdisciplinary)
Psychology (interdisciplinary)
Zoology

Ecology
Agriculture
Sociobiology
Chemistry
Geology
Geography (interdisciplinary)
Physics
Engineering (interdisciplinary)
Medicine (interdisciplinary)
Humanities (Social science)
History
Philosophy (interdisciplinary)
Communication (interdisciplinary)
Behavior
Sociology
Psychology (interdisciplinary)
Ethology (Animal behavior)
Sociobiology
Zoology
Agriculture
Geography (interdisciplinary)
Economics
Business
Engineering
Engineering is an interdisciplinary branch that overlaps
with other top-level categories.
Technology

The list of scholars/scientists in different fields, produced
by the world in different countries in different eras runs in
to thousands. A few to be named are: Plato-347 BC,
Socrates 399 BC, Aristotle- 322 BC, Galileo (1642 AD),
Archimedes- 212 BC, Ptolemy, Hippocrates, Sigmund
Freud- 1839, Pythagoras 500 BC, Euclid 300 BC,
Brahmagupta 670 AD, Al Khwarzmi 850 AD, Al-Kindi,
Al-Razes, Ibn-i-Rushd (Averros), Ibn-i-Sina (Avecina),
Aven Zoor (Ibn-i-Zoar), Al-Hazan (Abul Hasan), Al

Mamun, Ibn-i-Junus, Nasir-ud-Din Tusi, Albani, Al-Batan, Al-Bucasis of Cordova, Issac Newton 1727, Einistien etc.

A history of ingenuity:
Humans are an ingenious species. From the moment someone bashed a rock on the ground to make the first sharp-edged tool, to the development of Mars rovers and the Internet, several key advancements stand out as particularly revolutionary. Among innumerable inventions, these are our picks for the 10 most important inventions of all time.

The wheel: Before the invention of the wheel in 3500 B.C., humans were severely limited in how much stuff we could transport over land, and how far. Wheeled carts facilitated agriculture and commerce by enabling the transportation of goods to and from markets, as well as easing the burdens of people traveling great distances. Now, wheels are vital to our way of life, found in everything from clocks to vehicles to turbines.

The compass: Ancient mariners navigated by the stars, but that method didn't work during the day or on cloudy nights, and so it was unsafe to voyage far from land. The Chinese invented the first compass sometime between the 9th and 11th century; it was made of lodestone, a naturally-magnetized iron ore, the attractive properties of which they had been studying for centuries. Soon after, the technology passed to Europeans and Arabs through nautical contact. The compass enabled mariners to navigate safely far from land, increasing sea trade and contributing to the Age of Discovery.

The printing press: The German Johannes Gutenberg invented the printing press around 1440. Key to its development was the hand mold, a new molding technique that enabled the rapid creation of large quantities of metal movable type. Printing presses exponentially increased the speed with which book copies could be made, and thus they led to the rapid and widespread dissemination of knowledge for the first time in history. Twenty million volumes had been printed in Western Europe by 1500. Among other things, the printing press permitted wider access to the Bible, which in turn led to alternative interpretations, including that of Martin Luther, whose "95 Theses" a document printed by the hundred-thousand sparked the Protestant Reformation.

The internal combustion engine: In these engines, the combustion of a fuel releases a high-temperature gas, which, as it expands, applies a force to a piston, moving it. Thus, combustion engines convert chemical energy into mechanical work. Decades of engineering by many scientists went in to designing the internal combustion engine, which took its (essentially) modern form in the latter half of the 19th century. The engine ushered in the Industrial Age, as well as enabling the invention of a huge variety of machines, including modern cars and aircraft.

The telephone: Though several inventors did pioneering work on electronic voice transmission (many of whom later filed intellectual property lawsuits when telephone use exploded), Alexander Graham Bell was the first to be awarded a patent for the electric telephone in 1876. The invention quickly took off, and revolutionalized global business and communication.

The light bulb: When all you have is natural light, productivity is limited to daylight hours. Light bulbs changed the world by allowing us to be active at night.

According to historians, two dozen people were instrumental in inventing incandescent lamps throughout the 1800s; Thomas Edison is credited as the primary inventor because he created a completely functional lighting system, including a generator and wiring as well as a carbon-filament bulb. As well as initiating the introduction of electricity in homes throughout the Western world, this invention also had a rather unexpected consequence of **changing people's sleep patterns.** Instead of going to bed at nightfall (having nothing else to do) and sleeping in segments throughout the night separated by periods of wakefulness, we now stay up except for the 7 to 8 hours allotted for sleep, and, ideally, we sleep all in one go.

Penicillin: It's one of the most famous discovery stories in history. In 1928, the Scottish scientist Alexander Fleming noticed a bacteria-filled Petri dish in his laboratory with its lid accidentally ajar. The sample had become contaminated with a mold, and everywhere the mold was, the bacteria was dead. That antibiotic mold turned out to be the fungus Penicillium, and over the next two decades, chemists purified it and developed the drug Penicillin, which fights a huge number of bacterial infections in humans without harming the humans themselves. Penicillin was being mass produced and advertised by 1944. This poster attached to a curbside mailbox advised World War II servicemen to take the drug to rid themselves of venereal disease.

Contraceptives: Not only have birth control pills, condoms and other forms of contraception sparked a sexual revolution in the developed world by allowing men and women to have sex for leisure rather than procreation, they have also drastically reduced the average number of offspring per woman in countries where they are used. With fewer mouths to feed, modern families have achieved higher standards of living and can provide better for each child. Meanwhile, on the global scale, contraceptives are helping the human population gradually level off; our number will probably stabilize by the end of the century. Certain contraceptives, such as condoms, also curb the spread of sexually transmitted diseases. Natural and herbal contraception has been used for millennia. Condoms came into use in the 18th century, while the earliest oral contraceptive "the pill" was invented in the late 1930s by a chemist named Russell Marker.

The Internet: It really needs no introduction: The global system of interconnected computer networks known as the Internet is used by billions of people worldwide. Countless people helped develop it, but the person most often credited with its invention is the computer scientist Lawrence Roberts. In the 1960s, a team of computer scientists working for the U.S. Defense Department's ARPA (Advanced Research Projects Agency) built a communications network to connect the computers in the agency, called ARPANET. It used a method of data transmission called "packet switching" which Roberts, a member of the team, developed based on prior work of other computer scientists. ARPANET was the predecessor of the Internet.

Present State: Today information is a powerful tool. People are increasingly becoming dependent on information generation in the electronic media the world over. A user can now have all the latest information that he needs on his finger tips: electronic newspapers, yellow pages, telephone directories, stock exchange prices etc. Access to information as a basic right can stimulate the world's economy to the benefit of all. The business community has now come to understand information as a valuable commodity required for planning, directing, controlling, decision-making, motivating, and fore-casting and so on to ensure positive and gainful operation.

A quarter century ago about 50,000 computers existed in the whole world. Today there are more than 150 million. A typical American car today has more computing power than the lunar-landing craft had in 1969.

If computer is the most important thing that man invented since the wheel, software is the fuel that sets the wheels of the machine running.

In 1960 a transatlantic cable could carry only 138 conversations simultaneously. Today a fiber-optic cable carries 150 million. No communication has grown faster than the internet, which already connects more than 50 million users worldwide. Anybody with a computer modem and a telephone can Tele-shop, Tele-bank, and Tele-learn 24 hours a day.

The story of telecommunications can be traced to about two centuries back ranging from first facsimile (FAX) machine of 1843 to the Digital Cellular phone in 1990.

A report by the OCED estimates that more than half of the GDP in rich economies is now knowledge based, including industries such as Telecommunications, Education, Television, Computers, Software and Pharmaceuticals.

Abstract Knowledge has been the staple source of competitive advantage for many organizations for hundreds of years. During the 1990s, the onset of Internet and Information Superhighway, allowed KM to take off. It provided more opportunities for knowledge sharing and knowledge transfer than there had been in the past. This paper discusses the paradigm shift from agricultural to industrial economy and then to new Knowledge Economy. It provides a conceptual view of Knowledge management and its key drivers- highlights the evolution and the functions of portals- also elucidates different tools and technologies which act as platform to bring people together to share knowledge in the form of expertise, competencies, and skills irrespective of time and space constraints. It concludes that the future is for Knowledge Portals that provide flexible knowledge environment for large number of users.

Keywords : Knowledge Driven Era, Knowledge Portal Technologies

1.Introduction: Knowledge and innovation have played an important role in the development of society throughout history. The transformation from Agrarian to Industrial Society and now to the Information and Knowledge Society has largely been brought about as a result of the accumulation of Knowledge and the advances in Information and Communication Technologies. Digitization, open systems standards, and the development software and supporting technologies for the application of new computing and communication systems – including scanning and imaging technologies, memory and storage technologies display systems and copying technologies have intensified the move towards Knowledge codification, increased share of codified knowledge in the knowledge stock of advanced economies. All knowledge that can be codified and reduced to information can now be transmitted around the world.

2. *Paradigm Shift:* "There is a central difference between the old and the new economies: the old industrial economy was driven by economies of scale, the new information economy is driven by the economies of networks" – Carl Shapiro and Hal R. Verian – Information rules.

In an agricultural economy land is the key resource. In an industrial economy natural resource such as coal, iron ores are the main resources. A knowledge economy is one in which knowledge is the key resource. … One in which the generation and the exploitation of knowledge has come to play the pre dominant part in the creation of wealth. It is not simply about pushing back frontiers of knowledge; it is also about the more effective use and exploitation of all types of knowledge in all manners of economic activity. The knowledge economy is emerging from two defining forces; the rise in the knowledge intensity or economic activities, and the increasing globalization of economic affairs. The combined forces of information technology revolution and the increasing pace of technological change are driving the rise in knowledge intensity. Globalization is being driven by national and international deregulation, and by the IT related communication revolution.

3. *Knowledge Management*: Change is the order of the day. Increases in the organizational information and change have created a great need to manage knowledge to ensure

effectiveness. Knowledge management can be viewed as the process of identifying, organizing and managing knowledge resources. These include explicit knowledge (information), 'know how' (learning capacity), 'know who' (customer capacity) and tacit knowledge in the form of skills and competencies. Key drivers for knowledge management. Some of the key drivers for knowledge management are mentioned below:

• *Achieving organizational efficiency:* Knowledge management plays a significant role in achieving organizational efficiency. In the new economy, speed and responsiveness are determining success factors. Indeed, in the Internet world where customers expect services to be available on a 24-hour basis, firms have no choice but to make a quantum-leap improvement in various aspects of their services. This in turn has created the need for organizational to have organized information to facilitate their operations, information that is timely, accurate, useful and, more importantly, tailored to the organization's need.

• *Staying ahead of the competition*: In order to stay ahead of the competition, firms nowadays understand fully the need to know their customers and their competitors very well. Lee, Wee and Bambang-Walujo (1991) highlighted that intelligence gathering/market intelligence is a crucial activity that companies must undertake in today's competitive business world.

• *Maximizing Organizational potential:* The ability of an organization to innovate and create knowledge will depend largely on its ability to capture and manage knowledge. However, knowledge creation is an incremental process that requires the existence of a knowledge infrastructure. Knowledge management is about identifying and managing existing knowledge resources. It is also about making these resources available for knowledge workers to use in their work. Knowledge management professionals can play an important role in facilitating the knowledge creation process by facilitating knowledge-sharing and providing access to knowledge resources as and when these resources are needed.

• *Managing intellectual capital*: In the knowledge-based economy, the value of an organization is largely measured by the value of its knowledge (or intangible) assets. Intellectual capital involves human capital, customers' capital, structural capital and business intelligence capital. Each of these categories relies heavily on the creation and management of knowledge assets.

4. *Knowledge Portal Technologies:* The World Wide Web (WWW) has paved the way for the information age. With a competitive market demanding more information from various quarters, the Web has turned out to be a variable resource. In the early days Web surfers were frustrated by the delay in finding the information they needed. The first major information retrieval leap came from the development of Web search engines such as Lycos, Excite, AltaVista, etc. While everyone lauds the Web for offering unbridled opportunities to explore and discover new things, many still want someone else to aggregate a variety of interesting content in one place instead of creating massive and unwieldy bookmark files in their browser. These new online services 9 3 are Web sites, delivering the old formula of content, community and core services, but in a

new package and transformed as Web portals. Evolution of portals, Search Engines, Navigation Sites, Portals, Search Indices, Search Indices Content Management, Categorized Content, Collaboration Personalization Key Functions (Awad, Elias M. and Ghaziri, Hassan M.-2004) The main goal of a portal is to provide a single point of access to all information sources. Therefore portals must be the ultimate tools for universal integration of all enterprise applications. At the same time because every individual has different information needs and knowledge uses, portals have to deliver a personalized interface. Keeping in view the complexity of these challenges portals must include the following functionalities:

• *Gathering*: Documents created by knowledge workers are stored in a variety of locations. In order to be accessible data and documents need to be captured in a common repository.

• *Categorization*: This category profiles the information in the repository and organizes it in meaningful ways for navigating and searching. Portal should support categorization at all levels, including the knowledge worker and customer levels.

• *Distribution*: This facility supports the distribution of structured and unstructured information in the form of electronic or paper documents.

• *Publish*: This facility publishes information to a broader audience, including individuals outside the organization.

• *Personalization*: This is a key component of portal architectures because it allows individuals to enhance their productivity. It is becoming a necessity for successful portals. This is due to the proliferation of information available through the portal. To take advantage of this facility knowledge workers must be able to manage or prioritize the delivery of information on task function or interest basis.

• *Search/Navigate*: This component provides tools for identifying and accessing information. The knowledge worker can either browse or submit a query. Collaboration Knowledge portals provide a platform for people to engage in discussion and exchange information. The framework includes interactive facilities such as chat sessions, bulletin boards, and application sharing together with shared workspaces, whiteboards, and collaboration and authoring tools. Collaboration in the knowledge management context is the ability for two or more people to work together in a coordinated manner over time and space using electronic devices. One has to distinguish between two types of Collaborations:

• Asynchronous collaboration

• Synchronous Collaboration.

Below are given few advantages and disadvantages of Synchronous and Asynchronous Collaboration tools. Synchronous Collaboration, Asynchronous Collaboration, Teleconferencing, Electronic Mailing Lists. In use extensively by senior management Lists have been in use for a number of years and staff, conference telephone calls represent an extremely cost effective (if relatively expensive) collaboration technology.

Advantages: cheap technology, personal, immediate

Desadvantages: limited communication medium feedback, expensive, often does not work well across time zones, Computer Video/Teleconferencing, Web-Based Discussion Forums, Computer-based teleconferencing and there are a number of different online video- teleconferencing is a rapidly discussions forum applications in use.

Evolving technology that has tremendous Advantages : same as electronic mailing potential for distributed organizations lists except requires slightly faster Internet connection.

Disadvantages : cultural resistance, Web-Based Discussion Forums, Online Chat Forums, Lotus Notes Allow multiple users to communicate, Lotus notes is a comprehensive collaboration simultaneously by typing messages on tool that includes e-mail and groupware, a computer screen.

Advantages: comprehensive collaborative solution employing state-of-the-art technologies for communication, document management and work flow.

Disadvantages: expensive to deploy when compared with other collaboration technologies. Content Management. Another important issue handled by content management is the way documents are analyzed, stored and categorized. Once the documents have been gathered, they must be analyzed so that their content is available for retrieval and use by the system or end users. As documents enter the portal system, they are stored for later retrieval and display. However, it is not useful to simply put the documents away in their raw form. Systems typically analyze the documents content and store the results of that analysis so that subsequent use of the documents by the system and users will be more effective and efficient. As the number of documents under management grows, it becomes increasingly important to gather similar documents into smaller groups and to name the groups. This option is called categorizing.

The new technology trends in implementing portals: Portal & New Technology Directions Global, just-in-time knowledge sources and services, Analytic Tools, Intelligent Training, Collaborative Learning, Performance Support User-,task-,and situation tailored interaction, Human Computer Interaction, Collaborative Filtering Information, Brokers Knowledge, Integration Knowledge Management, Multimedia Multilingual Multi-document Digital Libraries, Seamless collaboration across geographic, temporal, organizational, and mission boundaries, Collaborative Environments, Intelligent agents to monitor, filter, search, extract, translate, fuse, mine, visualize and summarize information for a variety of operational needs, Intelligent Agents 9 6 Examples of Knowledge Portals (http://www.unesco.org) UNESCO has a vital role in gathering, transfer, dissemination and sharing of data, information and knowledge. UNESCO has created public domain portals for diverse groups of users with very rich contents.

Conclusion: Knowledge is the key source of a postindustrial society and telecommunication is the key technology. The advances in information and communication technologies, the Internet revolution, and the move towards the Information and Knowledge Society have highlighted the importance of knowledge and need for Knowledge management. Collaborative applications such as e-mail, calendaring, scheduling, shared folders and threaded discussions promote knowledge sharing and knowledge transfer. Both software vendors and knowledge-aware companies are investing huge sums in the development of efficient Knowledge Management solutions. These investments and the potentials of new technologies, additional bandwidth, and future Internet services will allow for a completely new form of process-oriented, user-centered portals that will cater for sophisticated users and provide knowledge for competitiveness.

Khwaja Baha-ud-Din Muhammad Naqshband (RA) (1317-1389AD)-a great Sufi Saint of Central Asia has classified knowledge in to three categories. One is bookish knowledge. Many books were written, the author died, the book was eaten by moths or perished in fire or floods and many forms of knowledge perished this way. This is grouped as perishable knowledge. The second form of knowledge is that of verification but this too is not a reliable one. Today we say sun is stationary, tomorrow we find it is in motion or atom is indivisible but next day we find it is otherwise. Thus this form of knowledge is unreliable based on theories. A third form of knowledge neither needs books nor verification, it is transferred from person to person and this is the most reliable one. You must consider that the scholar of this knowledge has reached its climax when he says, "I know nothing", as the ocean of knowledge has no boundaries and it is like a fathomless deep ocean. Then Khwaja says that I can contact a person who is thousands of kilometers away or who has passed away thousands of years ago. The former is known as "Tai Makan"; the later is known as "Tai Zaman" i.e. distance and time is no bar for him and surprisingly it is the lowest stage of this knowledge. This knowledge is called 'Ilim-i-Ludni'- Spiritual knowledge. From the above presentation it appears that the exoteric knowledge is in race with esoteric knowledge. Today we hear that research is in progress in electronic teleportation of physical bodies, besides travelling in to the past. A research scholar of Astrophysics of international repute stated that only 4 % of total space is known so far, besides the dark matter in space is still unexplored. There is no edge of space and nothing exists beyond that. We are yet to know what future knowledge has in store for us and what we consider impossible today may become a reality tomorrow. Here I would like to quote Dr. Iqbal:

عروجِ آدمِ خاکی سے انجم سہمے جاتے ہیں

کہ یہ ٹوٹا ہوا تارا مہِ کامل نہ بن جائے

With the progress of earthly human being, the stars are aghast, That this strewn star may not become a perfect moon.

Here I quote Shaikh Sadi Shirazi from his book Kareema
(Scrolls of Wisdom)

در فضیلت علم -- بیچ فضیلت علم کے

بنی آدم از علم یابد کمال -- بیٹا آدم کا علم سے پاتا ہے کمال

نہ از حشمت و جاہ و مال و منال -- نہ حشمت اور مرتبہ اور مال و اسباب سے

چو شمع ازپی علم باید گداخت -- مانند شمع کے واسطے علم کے چاہئے گھلنا

کہ بے علم نتوان خدا را شناخت -- کہ بغیر علم کے خدا کو نہ پہچان سکے

خرد مند باشد طلبگار علم -- عقلمند ہووے طلب گار علم کا

کہ گرم ست پیوستہ بازار علم -- کہ گرم ہے ہمیشہ بازار علم کا

کسی را کہ شد در ازل بختیار -- جو کہ ہو ازل میں نصیبہ ور

طلب کردن علم کرد اختیار -- طلب کرنا علم کا کیا اختیار

طلب کردن علم شد بر تو فرض -- طلب کرنا علم کا ہوا تجھ پر فرض

دگر واجب ست از پیش قطعہ ارض -- پھر واجب ہے واسطے اس کے کاٹ نا زمین کا

برو دامن علم گیر استوار -- جا دامن علم کا پکڑ مظبوط

کہ علمت رساند بدار القرار -- کہ علم تجھ کو پہنچائے گا بہشت میں

میاموز جز علم گر عاقلی -- نہ سیکھ سوا علم کے اگر عقلمند ہے تو

کہ بے علم بودن بود غافلی -- کہ بے علم رہنا ہووے غفلت

ترا علم در دین و دنیا تمام -- تجھ کو علم دین اور دنیا میں کافی ہے

کہ کار تو از علم گیرد نظام -- کہ کار تیرا علم سے لے آراستگی

ON THE EXCELLENCE OF LEARNING

Sons of Adam from learning will find perfection
Not from dignity, and rank, and wealth, and property.
Like a taper one must melt in pursuit of learning,
Since without learning one cannot know God.
A man of wisdom is a student of learning,
For the market of wisdom is always brisk.
Whoever is fortunate as regards Eternity,
Maketh choice of the pursuit of knowledge.
 This pursuit of knowledge is a duty on thy part,
Even if it be necessary to traverse the earth.
Go, seize fast hold of the skirt of knowledge,
For learning will convey thee to everlasting abodes.
Seek nought but knowledge if thou art wise,
For it is neglectful to remain without wisdom.
From learning there will come to thee perfection as regards
religion and the world,
For thine affairs will be settled by knowledge.
                    (Shaikh Sadi Shirazi RA) (1291 AD)

www.ingramcontent.com/pod-product-compliance
Lightning Source LLC
Chambersburg PA
CBHW081444170526
45166CB00008B/2311